SHOP INTERIOR SKETCH

SHOP INTERIOR SKETCH

2019년 1월 10일 발행

발 행	북이너스
주 소	경기도 고양시 일산 동구 고풍로 44-46
저 자	임성민
전 화	031-901-9058/9059
메 일	sungmin@paran.com
ISBN	979-11-6441-007-1
가 격	92,000원

• 파본 도서는 구매처에서 교환하여 드립니다.

GranitiFiandre spa   via Radici Nord, 112   42014 Castellarano (RE) Italy

1,72 mt

3 mt

1,50 mt

Think Big!

FIANDRE EXTRALITE

the ceramic "species" evolution

thickness 3 and 6 mm
in the innovative size 3x1,5 mt.

www.granitifiandre.com

Table of Contents

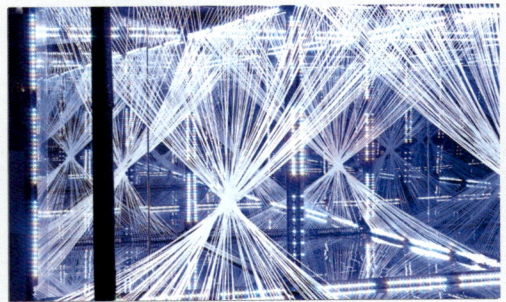

008–041

Retail
Omonia Bakery
Sneakerology
The OPUS Shop
52
Flavaboom
Kirk Originals Flagship Store
Shoesme
Rolls
2D3D Chairs for Issey Miyake
Hermès Rive Gauche
ARTIFACTS Nanshi Boutique
ARTIFACTS Dunhua Boutique

042–097

Restaurant/Bar
Cienna Ultra Lounge
Tree Restaurant
Tang Palace, Hangzhou
Everyday Chaa
Niseko Village Look out Café
Happo-en the Hakuhou-kan
Twenty Five Lusk
Santa Rita Restaurant
M.N.Roy Club
Zebar a Live Bar in Shanghai
Phill
Wienerwald
Holyfields Frankfurt
smith&hsu Teahouse
+Green
What Happens When

098–139

Office
No Picnic
Obscura Digital Headquarters
Bajaj Corporate Office In India
Sugamo Shinkin Bank: Tokiwadai Branch
Sugamo Shinkin Bank: Shimura Branch
De Nije Gritenije
Open Lounge
Thin Office
Studio SC
Televisor
Tribal DDB Amsterdam
Rabobank Nederland
DDB in Singapore

140–151

Hotel
Miura Hotel
The Club

156–199

Exhibition/Showroom

Brunner Salone Internazionale del Mobile 2011
The Design Bar
GE Healthymagination Showcase
The Orange Cube and RBC
Chopin's Visiting Card
Exhibition in the Poland Pavilion for Expo 2010 in Shanghai
Prague Quadriennale 2011
Decameron
Docks en Seine
Zaha Hadid: Form in Motion
Zaha Hadid, une Architecture in Mobile Pavilion
Roca London Gallery

151–155

Health/Beauty

Hairu Hair Treatment

200–231

Features

Music Hall Eindhoven
Big Cheese
More Fun for Your Life and Work

232–255

Designer— Streamline Icon

Metropol Parasol
Danfoss Universe
An der Alster 1
Level Green
M.over.Wall
Pre.text / Vor.wand
Dupli.Casa

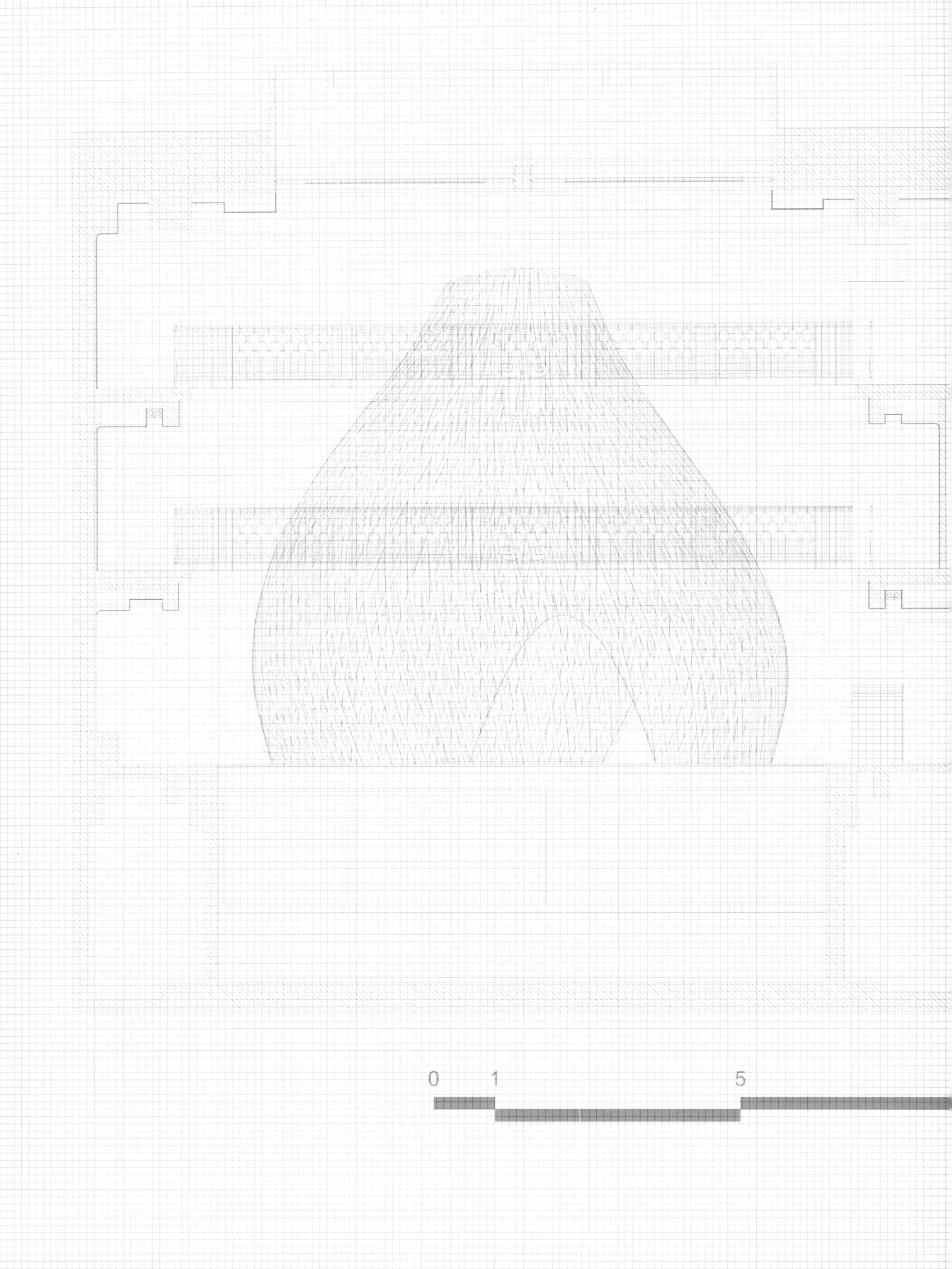

RETAIL

_01

OMONIA BAKERY

_bluarch architecture + interiors + lighting

This bakery is a brand new project for the family behind the renowned Omonia brand famous for its Greek pastries. It sells pastries and breads prepared on premises in the see-through kitchen.

The design of this store celebrates indulgence... the suspension of one's everyday grind through the consumption of a sweet delight. The space is soft and warm... sexy and decadent... as chocolate.

Much like the physiognomy of a pastry, this design wants to offer the exciting anticipation of a pastry in-fieri... the liquid concoction, the minced ingredients... The space shifts organically with the narrative of flavors as patrons taste the succulent goods.

The main feature of the 1,000SF interior space is a fluid surface (clad with ¼ chocolate brown Bisazza tiles) which covers the ceiling and the side walls to different heights. This surface warps in bubbles and negotiates a system of 6-inch tubular incandescent light bulbs... and an arrangement of red cedar wood spheres. The epoxy flooring continues to the walls via filleted corners. A shelf and LED strips navigate the transition with the chocolate surface.

The kitchen is exhibited to the public, as it sits simply within a tempered glass box... Therefore, the exquisite level of craftsmanship of the project (with its unforgiving alignments and complex details) is paralleled with the refined artisanship of Omonia's pastries.

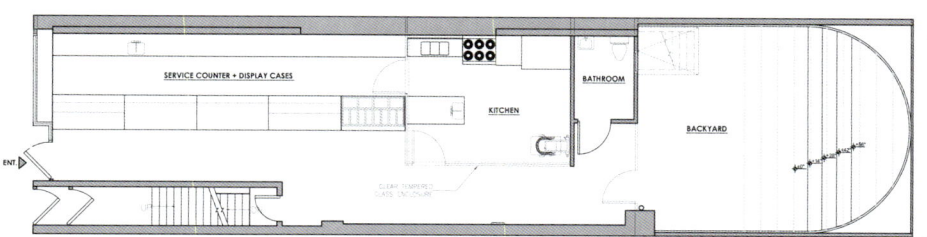

_FLOOR PLAN

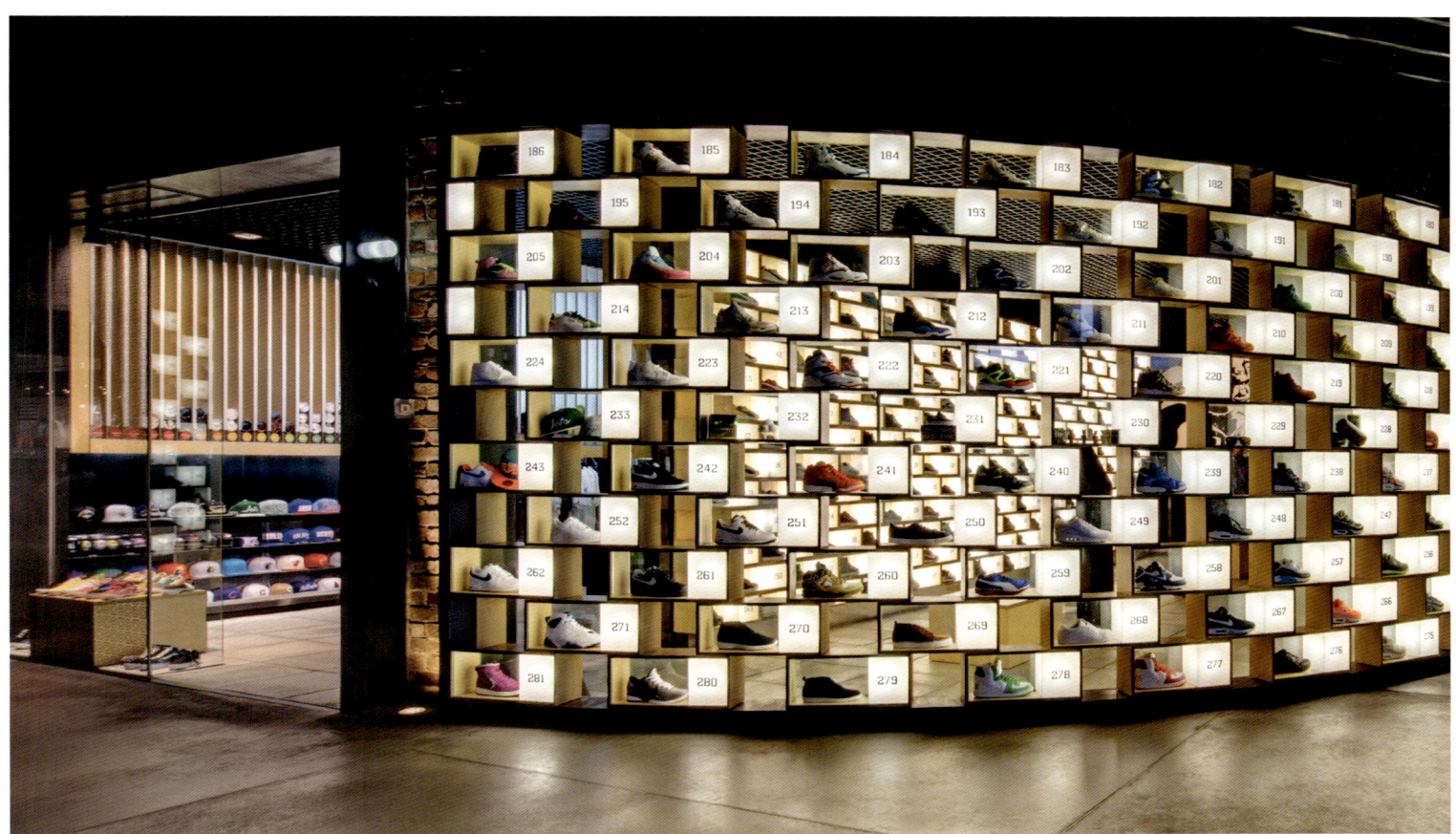

SNEAKEROLOGY

_Facet Studio

In each of the 200mm x 600mm boxes, one by one, sneakers are carefully collected. The boxes are repeated, and offset by half unit on each level, and carried through repeatedly over an entire wall. Something which has little meaning on its own, when repeated 281 times over, creates an euphoric effect for one to experience a heightened emotion. The merchandises are neatly displayed in a fashion similar to the museum artefacts. Through touch panels centrally located within the shop, one can gain further understanding of the background stories of the merchandises. Although there is really no such field of study as 'sneaker-ology', by placing our design focus on ways to correctly understanding the merchandises, it is for us an attempt at capturing 'sneakers' in a scholarly fashion.

'That one is nice... this one is nice too!' There is no better way to shop than whilst enjoying an academic high.

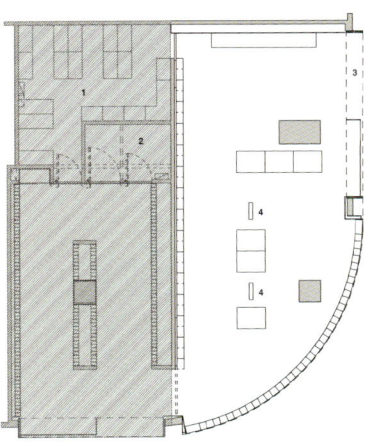

Sneakerology Floor Plan

1. STORAGE
2. CHANGING ROOM
3. ENTRY
4. INTERACTIVE PANEL

_FLOOR PLAN

Client_ Sneakerology   Country_ Australia   Photography_ Katherine Lu   Design_ Facet Studio

THE OPUS SHOP

_Paradox Studio

It's the world's one and only purse hanger specialty shop by Paradox Studio. OPUS is a brand specializing in purse hangers, which can be placed securely on the edge of the table to hang your purse, hence free up space at the table and on the seats, and free up your hands for more activities. OPUS Taipei is the first shop for the brand and was designed to be a multi-purpose space that can be used for meetings, product launches as well as a retail store.

OPUS Taipei is located in the city's fashion district. The previous use for this location was a garage and the space was converted into a small storefront during the economic recession.

The store is merely 2.3m wide and 4.5m long, which is about 10.5 m² and is a very petite space. To overcome the size limitation of the store, we designed a perspective illusion by painting yellow colour blocks (using OPUS' signature colour) on white walls to create the impression of a deeper and wider space. The rhythmic yellow blocks run along the two opposite walls of the store and converged into a horizontal line on the back wall which is highlighted with a clock custom-designed by us.

RETAIL | **Design**_ Paradox Studio | **Photography**_ Benjamin Chou | **Region**_ Taiwan | **Client**_ OPUS International

_THE OPUS SHOP   _RETAIL   _13

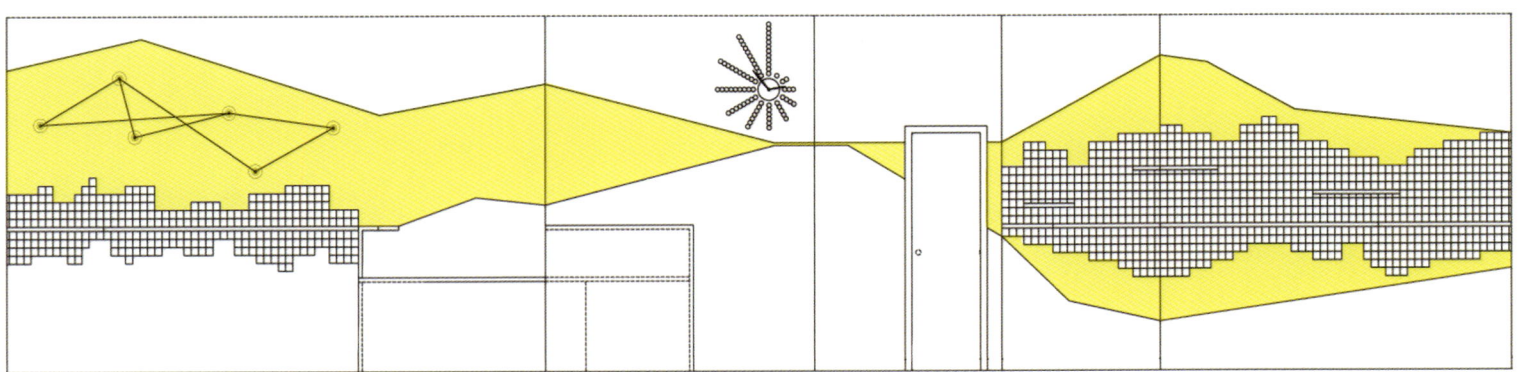

_ELEVATION

52

_Suppose design office
| Makoto Tanijiri

We had been requested to design a clothing shop in Shizuoka-shi Japan. In residential projects, we think about the relationship between the internal and external space. But for this project, we started to think about the relationship between the products and the two different spaces.

In the west there are many galleries that do not use spot lights but rather use natural light to light up the space. The reasoning for the use of natural lighting is that most paintings are painted under natural lighting and unless the painting is viewed under the state it was painted the true beauty of the painting will not show.

Could we not think the same for clothes? By creating a room that is like the outside and creating a room that is like the inside, the clothes, shoes and accessories can be placed in their rightful space.

A 9mm metal sheet wall was placed in a zigzag manner to separate the two different spaces and created big openings.

The space where light pours in from the skylight would be for outerwear, shoes and other products that would be used outside. The space that is lit up with warm artificial lighting would be for innerwear and stationary. Each product had it place and we placed them to the rightful place.

By creating an internal space and external space in a building using only natural light effect, we were able to find a new relationship between outdoor and indoor space.

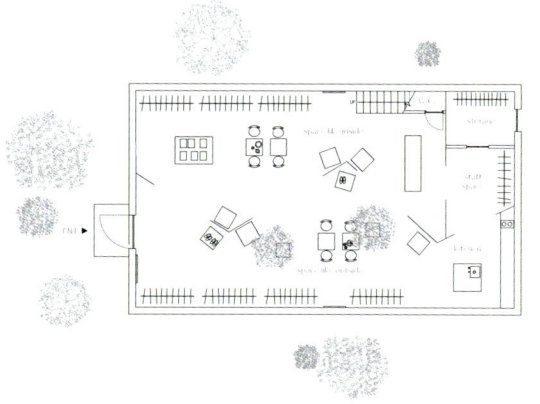
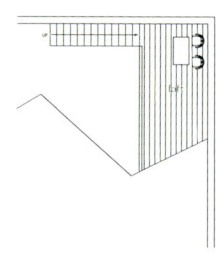
_PLAN
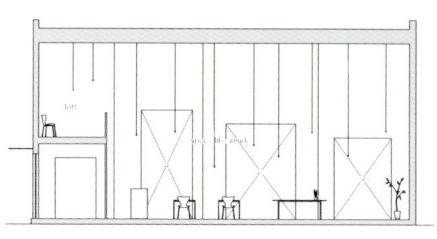
_SECTION

RETAIL Design_ Suppose design office | Makoto Tanijiri **Photography**_ Toshiyuki Yano **Country**_ Japan **Client**_ Mitsuko Terada

FLAVABOOM

_Dune

Flavaboom, the newest and coolest froyo brand just launched their premier self-serve frozen yogurt shop.
The futuristic, laboratory-like 1,500sqft state-of-the-art facility is located in the heart of Manhattan's prominent Chelsea neighborhood at 600 6th Avenue between 17th and 18th Street. Guests are welcome to sit back and enjoy their froyo creations while listening to great tunes.
The cutting-edge space features a sleek 30ft toppings bar, polished white walls and chic contemporary furniture outfitted by NY based eco-friendly and local manufacturing design company Dune.
Dune envisioned the concept in a yummy colour palette with glossy finishes and a vibrant logo. 'We handled the design and production of the graphics, architecture, design, millwork, and furniture. This prototype shop will be the world's first quick service food establishment to implement such a slick, progressive point of view.' says Richard Shemtov, Dune's president and founder.
One of the most exciting design elements at Flavaboom is the self-serve yogurt machine wall which is surrounded by an innovative, LED lighting system encased in handmade acrylic diffusers. This hi-tech backdrop will produce hundreds of light shows and can be seen from passersby outside.
The focus of Flavaboom is healthier food made from quality and natural ingredients. In addition to yogurt they are introducing an organic espresso bar and fresh baked goods. Guests can also enjoy the interactive fun-factor as they design their own custom frozen yogurt creations with 10 tasty froyo flavas and over 40 toppings. The yogurt is available in non-fat, low-fat, and vegan options; all is gluten free and kosher.
Flavaboom will no doubt become a destination among froyo and coffee lovers, the health conscious and design enthusiast alike.

_SELF-SERVICE YOGURT MACHINE **WALL**

KIRK ORIGINALS FLAGSHIP STORE

_Campaign

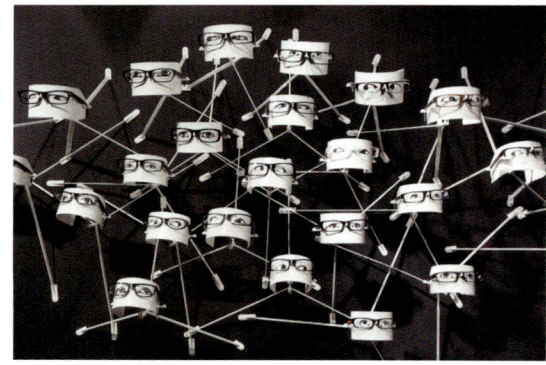

Campaign has created a dramatic interior for the London flagship store of global eyewear brand Kirk Originals, in Conduit Street, in the heart of London's west end, now officially open. A convivial space has been designed to convey the brand's heritage, ethos and be conducive to browsing and trying on the handcrafted frames on display, with full eye examinations and fittings available in the basement.

Taking inspiration from the brand's latest Kinetic collections, the flagship store design features displays of winking eyes in various guises. A series of larger than life lenticular printed eyes are suspended in the front window, meanwhile, a wall display of human-like 'winkie' runs the length of the store.

The 'winkies', 187 white powder-coated sculptural heads, each wear a unique frame and can be tilted and re-positioned to create clusters of onlooking craning heads. A restricted palette of monochromatic colours and modest materials including blue-grey painted walls and a dark grey floor keeps the spotlight firmly on the 'winkies' adorned with frames as if displaying works of art.

Integral to the shopping experience, the Kirk Originals identity is interwoven throughout the space: a succinct account of the brand's origins has been rendered in graphic text over two walls at the entrance; meanwhile a black and white projection on the back wall playfully reworks the Kirk Orginals logo through a continual kaleidoscopic loop.

_KIRK ORIGINALS **FLAGSHIP STORE** _RETAIL _19

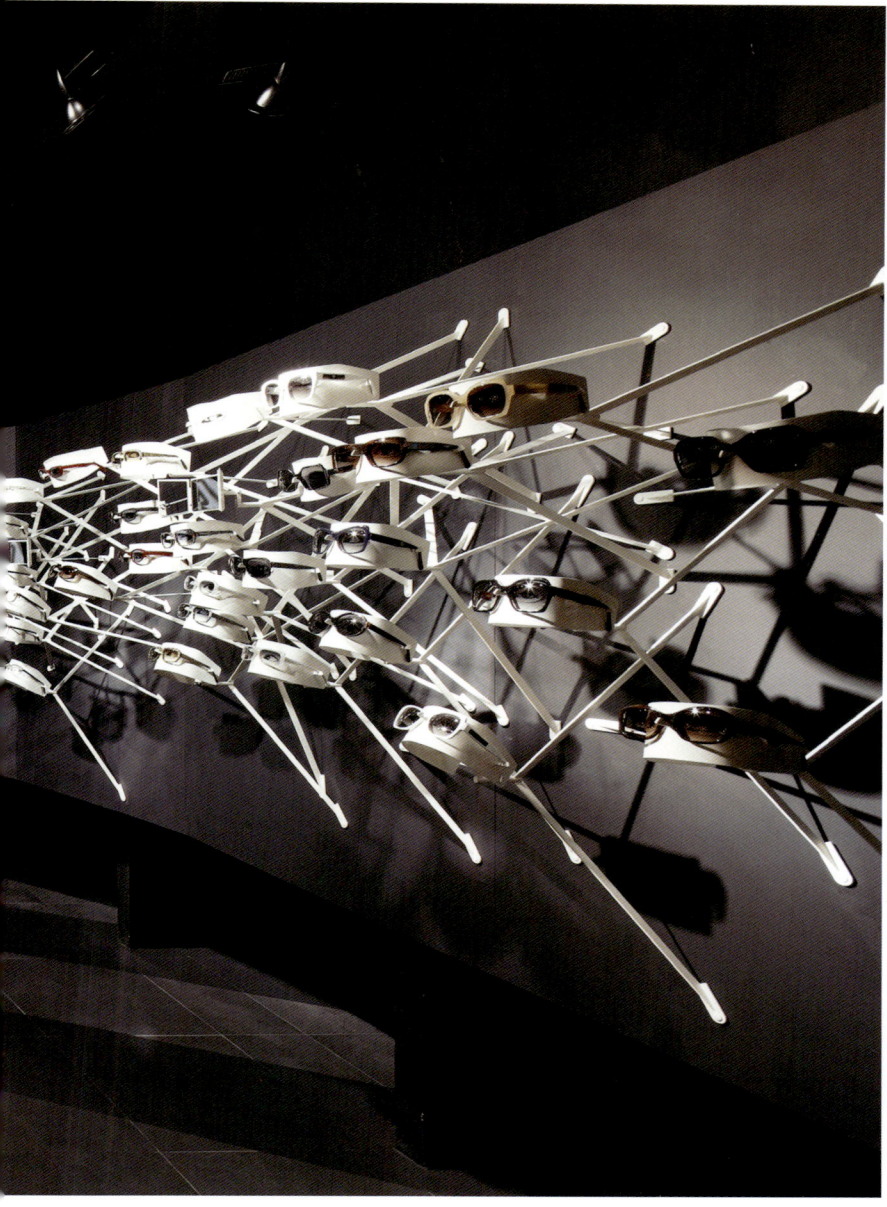

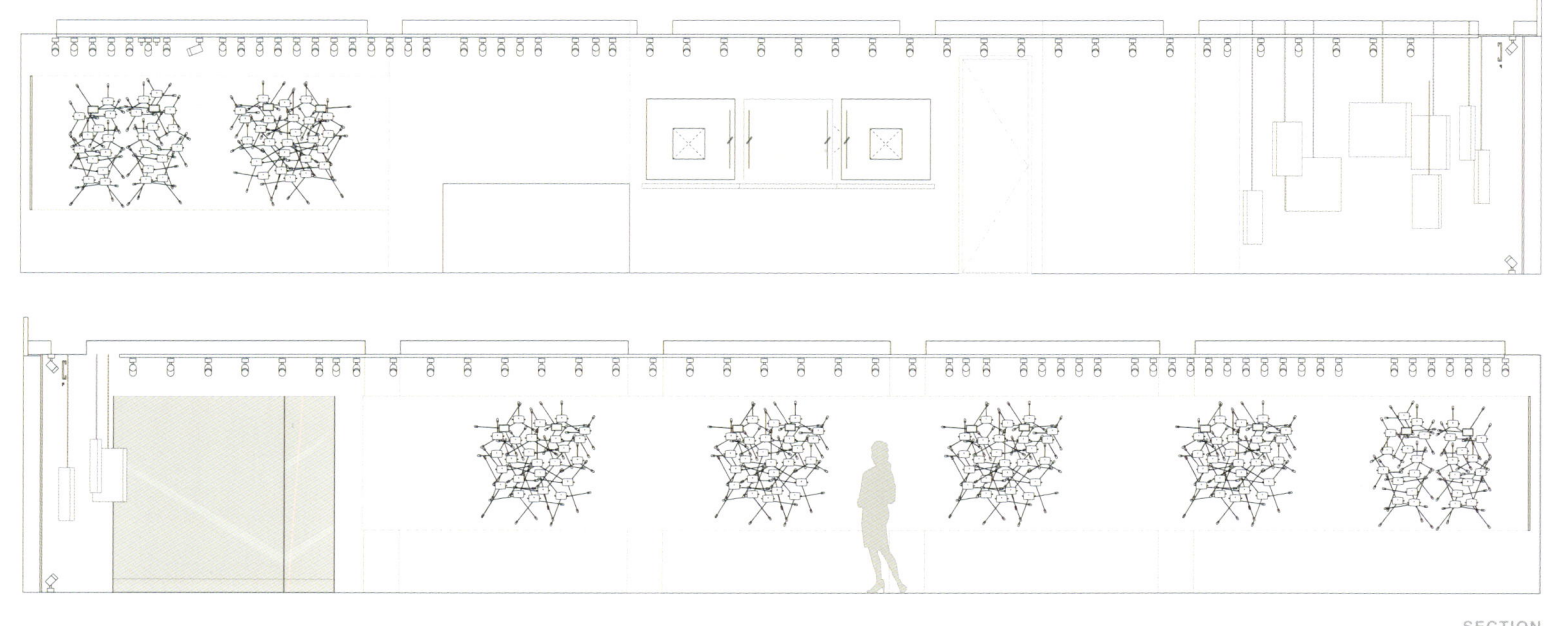

SECTION

Client_ Kirk Originals **Country_** UK **Photography_** Hufton + Crow **Design_** Campaign **RETAIL**

SHOESME

_CONCEPT

_Teun Fleskens

Shoesme is a high quality shoe brand for children, which handles the entire supply chain by itself – from production till distribution. The assignment was to create an interior with a possibility for later expansion. We asked ourselves the question if we could design a modular system in which you can build an interior, that is changeable, has the possibility to transform into a stand, and functions as the promotion material. Through this reasoning the concept was born.

We searched for inspiration in toys, especially building blocks. The shape of the die stands for playfulness and change. They are made from polyetheen and are rotation moulded (nontoxic and fully recyclable). The size originates from the function of the die. One die is a seat (children's height), two stacked dice create the height of a table and three dice are bar height.

After extensive experimenting, we invented a construction that makes it possible to stack and connect the dice. All elements have got four holes on each side; beech wooden pins fit these holes so the dice can be connected to one another. The tabletops are made of plywood and also contain the holes. In this way, Shoesme employees have the possibility to change the interior themselves, and also easily build a stand. As for branding, at the ending of the fair, Shoesme can give away parts of their stand, piece by piece. With this, the customers receive a free seat, which shows the Shoesme brand name. It also saves transport costs.

RETAIL | **Design_** Teun Fleskens | **Photography_** Michiel Maessen | **Country_** The Netherlands | **Client_** SHOESME International B.V.

The children's shoes are presented to retailers. Therefore, it is very important that the shoe is displayed as realistic as possible. We have designed lamps that can be hanged with a 360 degree radius. Every lamp contains a 4000 Kelvin cfl lamp, to display the colour exactly as it is. The inspiration for the lamps originates from the industrial look of the building. The lamps are designed in such a way that every wire has got a function. Through this, the wire became the lamp and becomes one with the industrial ceiling. Together, it creates a flexible, industrial but playful interior with additional value.
Another installation which makes the interior more playful is designed especially for children: when parents buy new shoes, their children will be given chance to pull a handle, which triggers a marble to make a trip for thirty meters on a set of metal tracks before dropping into a hole in the centre of a round plate. In a commercial way, children are supposed to recall the marble track play and ask their parents to patronize Shoesme shop again once the shoes are worn out. It is also nice in an aesthetic way.
This takes place at CAST: Centre for Accessories, shoes and bags. As a wholesaler centre, the CAST is unique. CAST is not just a permanent fair, but at the same time a place for inspiration. CAST contains suppliers in footwear, travel, and leather branches.

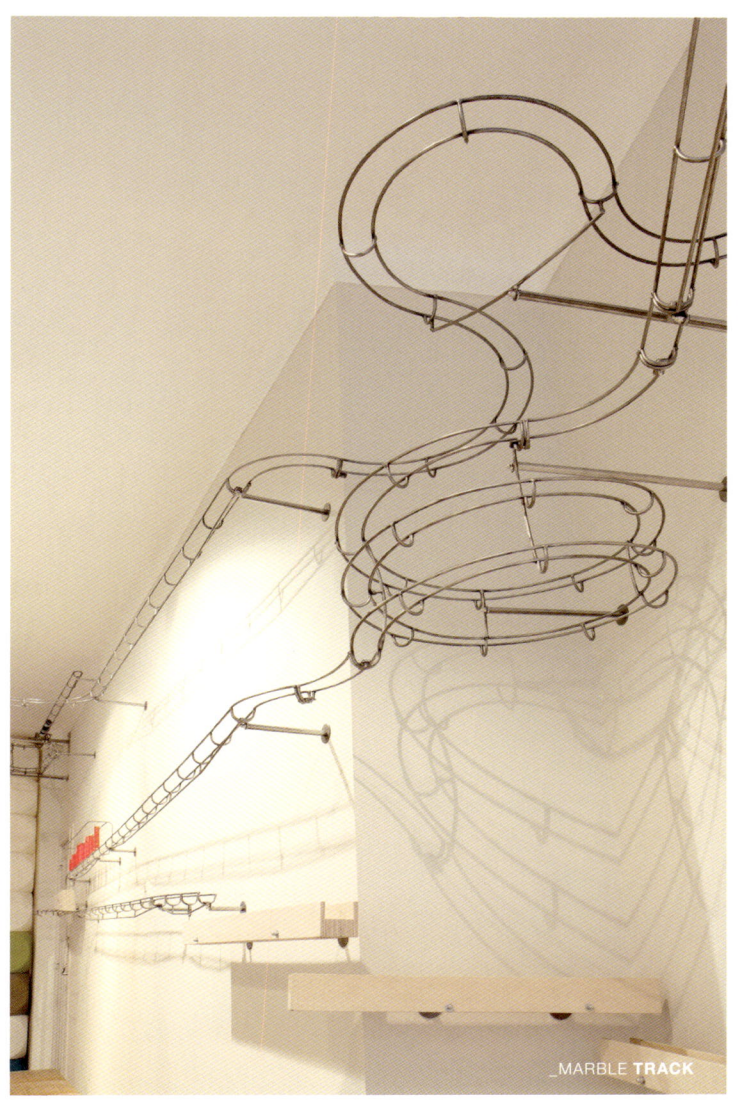

_MARBLE **TRACK**

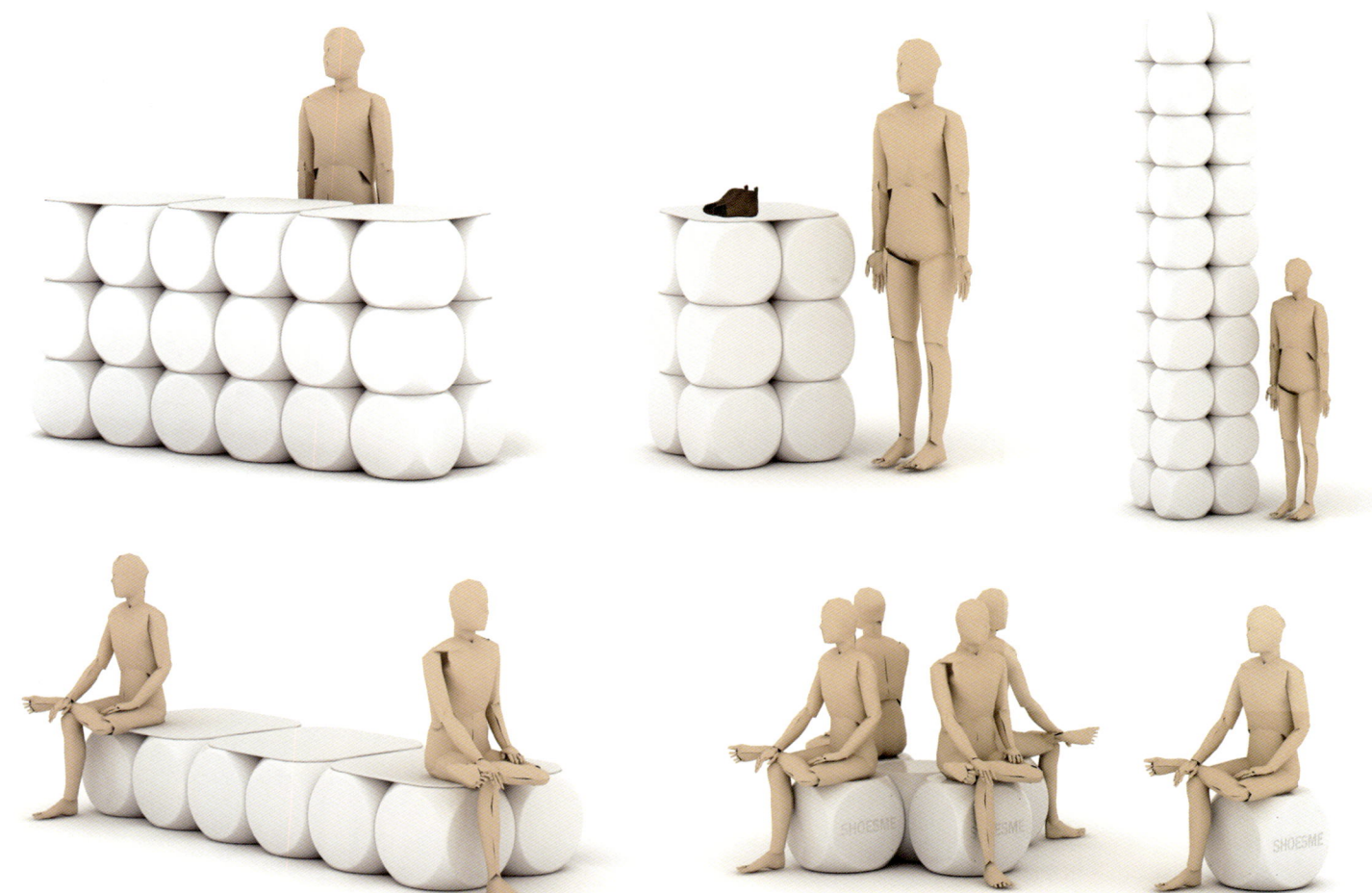

_CONCEPT

_RENDERING

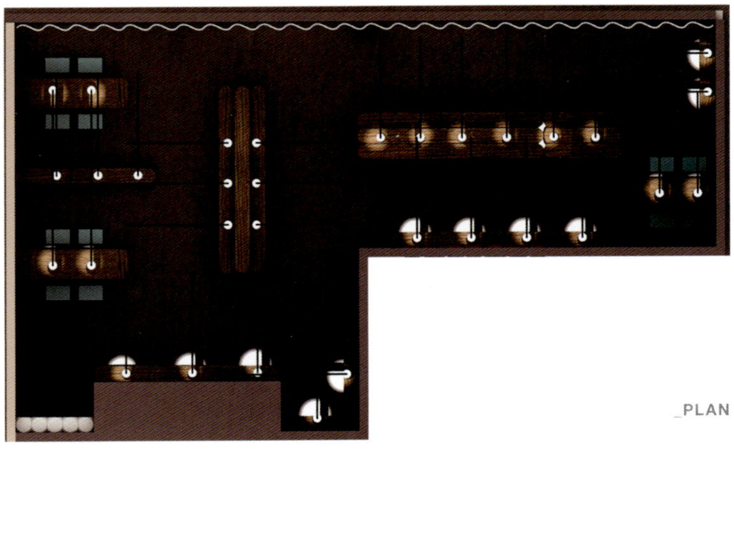

_PLAN

ROLLS

_Chikara Ohno / sinato

An installation in the shop space of Diesel Denim Gallery Aoyama in Tokyo.
The characteristic of the material used for the installation, which is aluminum, is that it is very thin and easily bent by hands, yet harder than cloth or paper. Therefore it possesses both soft and hard qualities.
By winding and sometimes extending this single, long strip of aluminum from the entrance to the back-end of the room, it creates a beautiful waving form, changing its function and features as the material strength changes.
The flexible quality of the material represents a loose interwining of the softness of clothes and hardness of architecture.

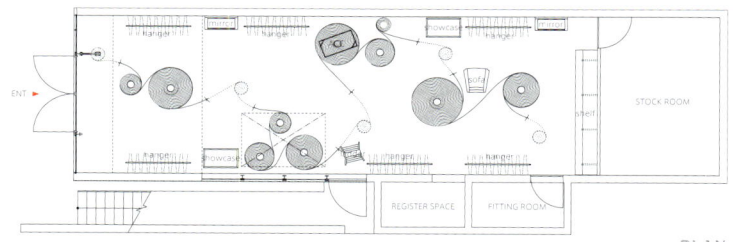

PLAN

RETAIL | Design_ Chikara Ohno / sinato | Photography_ Toshiyuki Yano | Country_ Japan | Client_ Diesel Japan Co.,Ltd.

2D3D CHAIRS FOR ISSEY MIYAKE

_Yoichi Yamamoto Architects

This project is a window display installation for the street front store. The backs of the chairs stand up from the stage, while the legs of the chairs are drawings on the floor of the stage.
If you look at the installation from one point in front of the shop window, the back of the chairs, which are three-dimensional objects, and the legs of the chairs, which are two-dimensional drawings, meet and create a single figure.
We expressed Issey Miyake's 'from 2D cloth to 3D dress' philosophy in our installation.

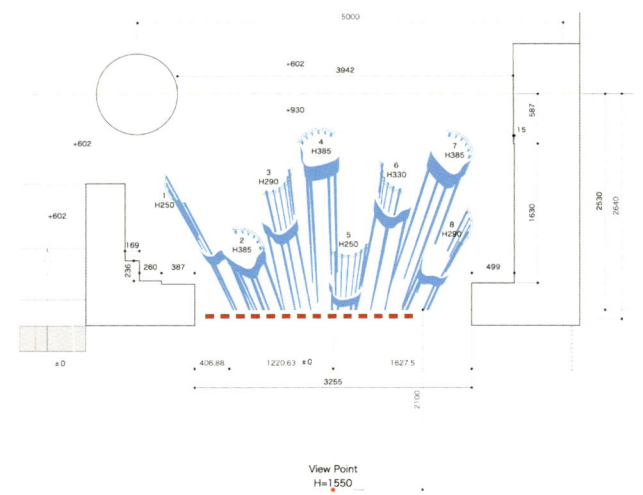

_PLAN

_2D3D CHAIRS FOR **ISSEY MIYAKE** _RETAIL _29

_VIEW FROM VIEW **POINT**

Client_ ELTTOB TEP ISSEY MIYAKE / GINZA **Country**_ Japan **Photography**_ Yoichi Yamamamoto Architects **Design**_ Yoichi Yamamoto Architects **RETAIL**

HERMÈS RIVE GAUCHE

_RDAI

Hermès has entrusted the RDAI agency, which is responsible for designing all the Hermès stores worldwide, with the design of a new space, singular and unexpected in Paris, a shop in a swimming pool – the Lutétia swimming pool, in the heart of the Saint-Germaindes-Prés quarter of Paris, has metamorphosed into the first Hermès boutique on the Left Bank. The architectural project led by Denis Montel and the teams at RDAI mixes contrasts and complementarities. It was imagined more in terms of volume than surface area, in m3 more than in m2.

Listed as a Historic Monument since 2005, the swimming pool built in 1935 has a strong architectonic character and a compelling identity, that of Art Deco – it is in the spirit of its age. After its closure, the swimming pool underwent varied and diverse uses and was transformed. The challenge was to translate some of the values intrinsic to Hermès into space: heritage and modernity, savoir-faire and creation.

The project has a double aim. First of all to respect, conserve and reinterpret the architecture of the swimming pool. The only important modification was the covering of the pool by means of concrete composite floor slab supported by a light structure. Underneath, the pool has been integrally preserved. The façade giving onto the rue de Sèvres, has kept its original appearance.

Then, to tell another story, one that is resolutely contemporary. This takes form through the appearance of three monumental ash huts which both disrupt the existing volumes and converse with them. The invasion of what was once the pool by these huts, flexible, light and nomadic, suggests the creation of houses within the house.

The shimmering of the water that was once here is evoked in a subtle way in the tones of the mosaics, in the effects of the lights…

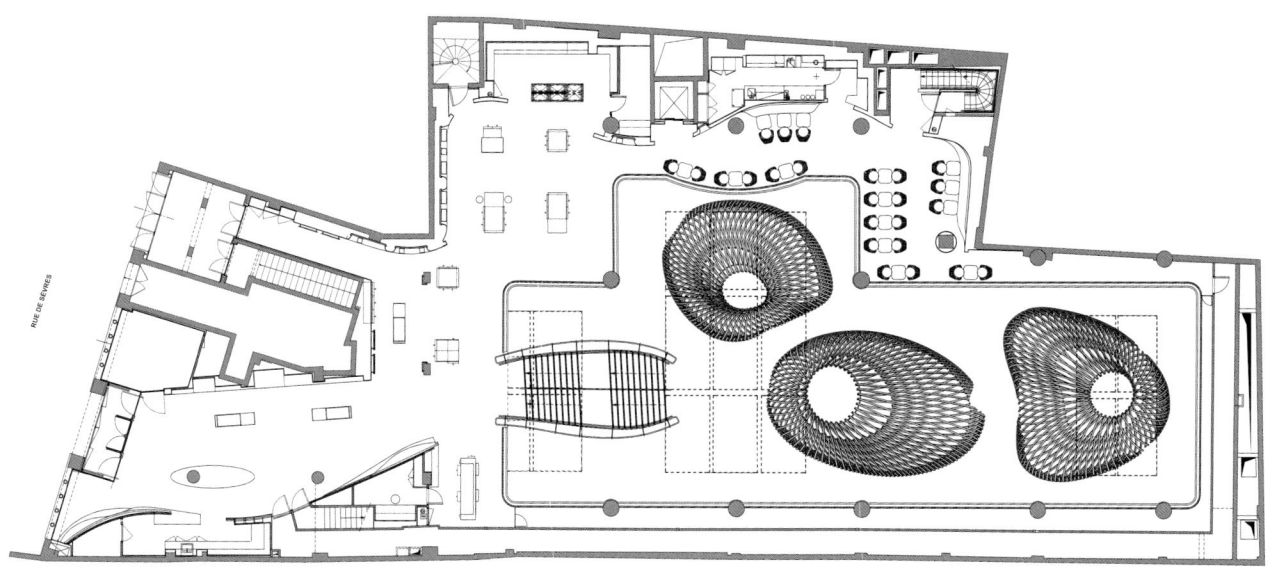

_GROUND FLOOR **PLAN**

Design_ RDAI **Photography**_ Michel Danance **Country**_ France **Client**_ Hermès Sellier

The Entrance
At the foot of an elegant apartment building from the mid 1930s, the façade of the Hermès store is discreet. An entrance portico in the centre between two windows, the entrance is like a lightwell overturned, horizontal, which attracts one irrevocably towards the light at the back, functions like a delicious trap into which the visitor lets himself slide, from crossing the threshold of the doors on the street until he reaches the swimming pool and its strange inhabitants, the huts. The lightly inclined ceiling, the walls curved and leaning inwards, covered with oak laths that leave recesses open as if floating in matter.

The Huts

Four pavilions with an organic design, in which some will recognise familiar forms from the plant or animal world, or from childhood... Others will liken these huts to the nests of tisserin birds. These pavilions of different form and dimensions are constructed in ash wood. They are self-supporting structures that rest on a system of woven wooden laths (profile 6x4cm) with a double radius of curves. The documentation and three-dimensional drawing of the complex geometry of each hut was made possible by the computer script written for each one of them. Rising to more than 9m in height, they lean progressively, as if attracted by the skylights. The huts house the Hermès collections. They seem to have simply alighted on the ground, lending the project its nomadic dimension. The fourth hut, which appears to be lying down, lines the staircase that naturally leads the visitor towards the pool and forms the link between the entrance and the open space of the swimming pool.

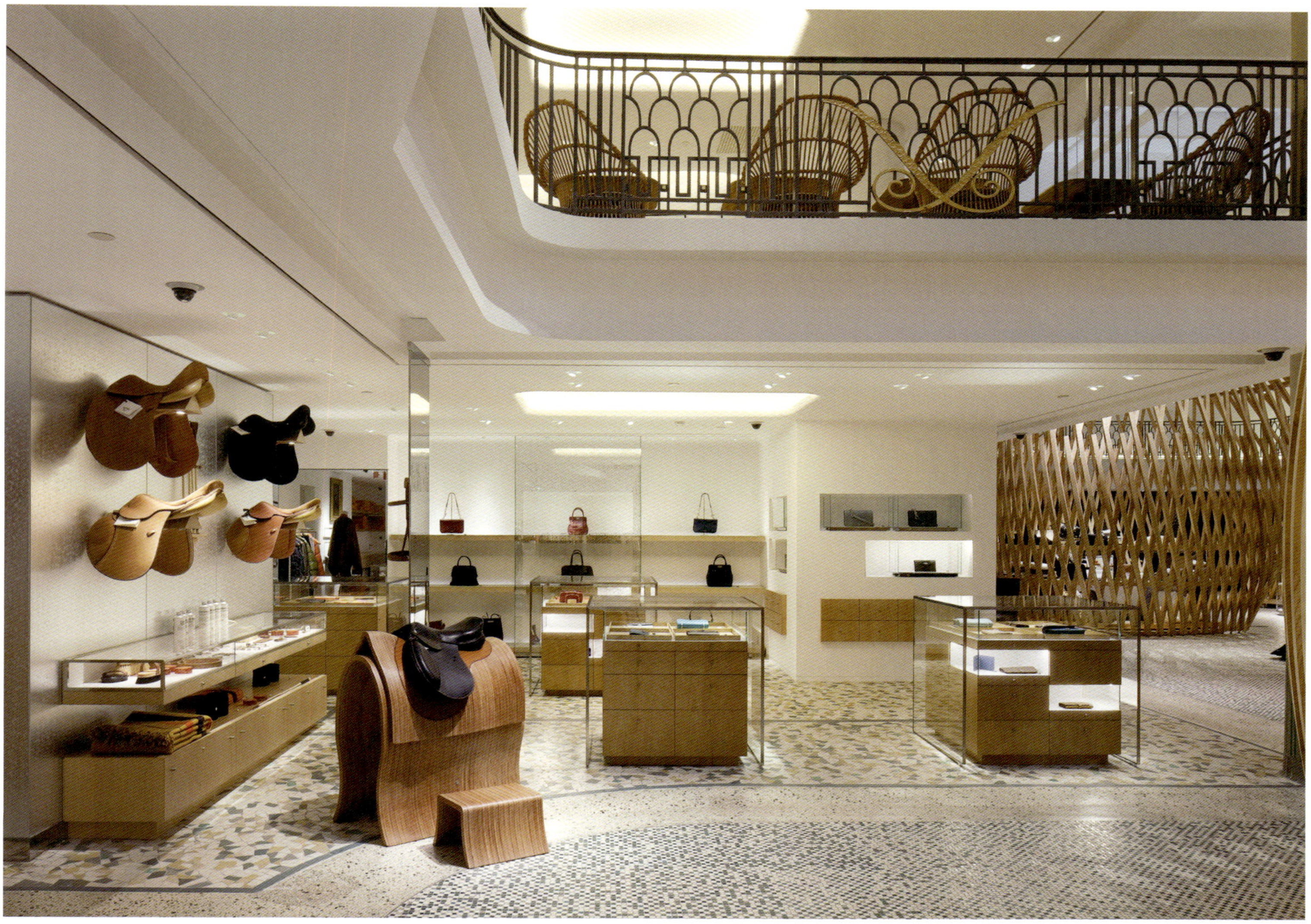

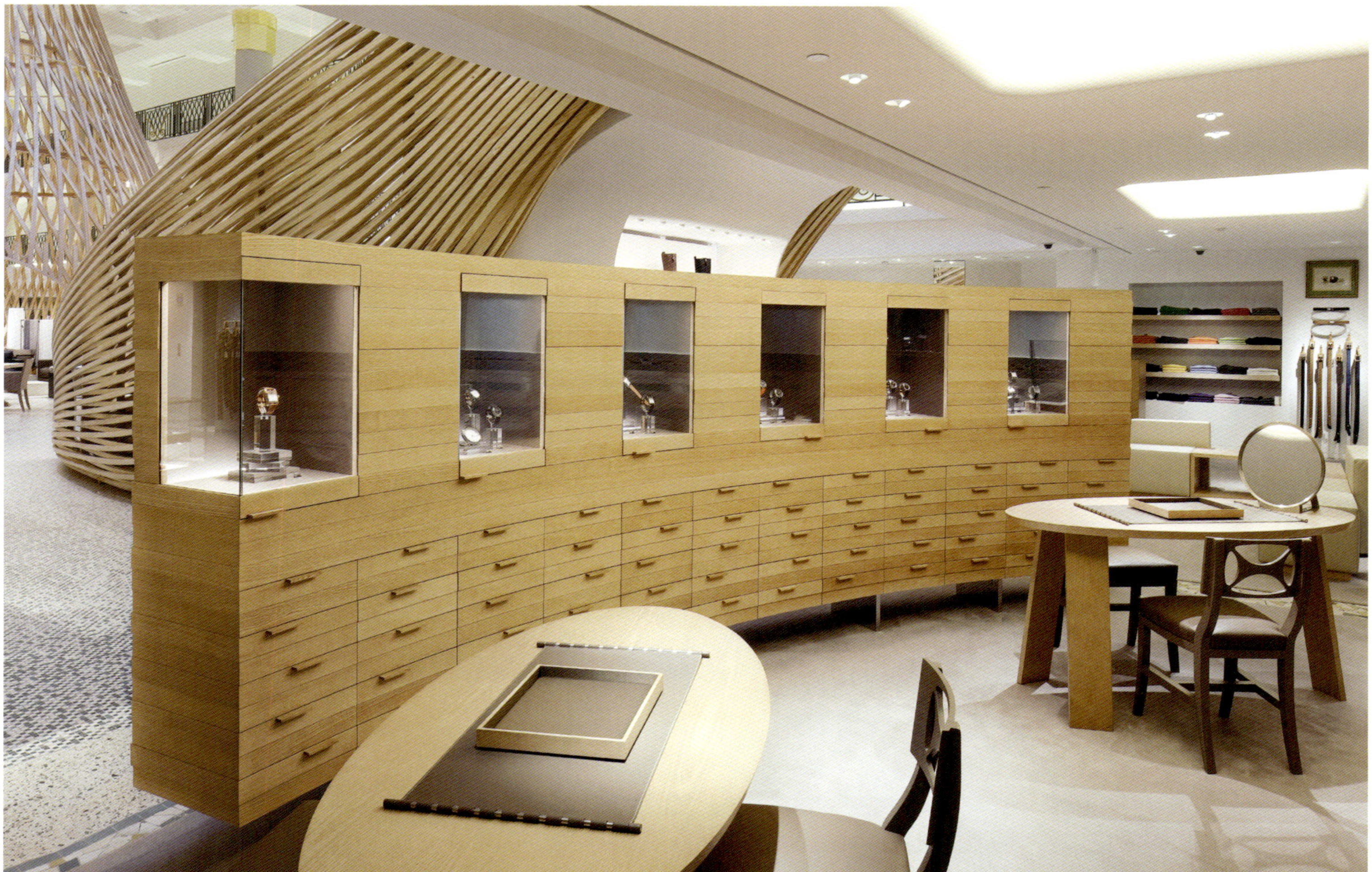

The Lighting
In such a volume, the lighting is crucial. The entire space is bathed in natural light that penetrates through the three large skylights above the atrium, softened only by a metal screen. At night the skylights are lit to avoid a 'black hole' effect. In order to avoid putting the spaces overlooking the pool that previously housed the changing rooms, in the shade, the effects had to be measured out, the contrasts that would otherwise have been too harsh attenuated. All the vertical panels are therefore also lightly illuminated. The undulating walls in white plaster, running around the ground floor, are lit from above by LED tape with the light source hidden from view. Lit from the interior, the huts appear as giant lanterns. A lighting device embedded in the floor, illuminates their vaults of latticed wood. Each hut has a large chandelier composed of a double ring of suspended wood. The shelving is lit by integrated and invisible LED tape.

The Mosaics
The Lutétia swimming pool is a mineral world. The floors, the columns, the staircases are covered in mosaics, broken tiles or granito. The existing ornamental elements on the floor and the walls have been preserved and restored. In the entrance to the store, a mosaic carpet with a Greek motif (a nod to the flooring of the Hermès store at 24 Faubourg Saint-Honoré) welcomes visitors. Following this desire for coherence, the steps and risers of the large newly created staircase are in granito.

WORK*shop* _ISSUE.TWO

ARTIFACTS NANSHI BOUTIQUE

_Straight Square Design

Taiwan based design firm Straight Square Design has recently completed two fashion retail stores in Taipei. The stores sell clothing, lifestyle accessories, books, and music.

The uses of black and white colours, floor to ceiling rotating glass doors, and three feet tall lifting clothes racks have created an unique shopping experience in a surreal setting, intentionally conveying the idea of futuristic shopping gallery in an achromatic space.

Nine rotating glass doors at the entrance turn 360 degrees in all directions have given the flexibility to the display window. The letters of 'ARTIFACTS' that are hung three feet height on each glass panel with a large scale have also given the attention to the passing customers.

The design takes advantage of the space with four feet height by filling the walls with repeated lifting clothes racks to increases the display area, forming a geometric pattern in the space. Reflection effect is another design element such as the shoes and the fitting room area, which have also given the repeated geometric pattern.

As the philosophy of ARTIFACTS, the design elements convey the idea of being rational, technological, and geometrical, giving the customers a surreal shopping experience.

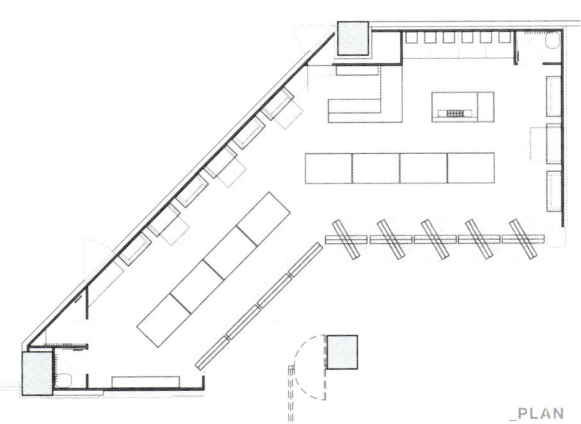
_PLAN

RETAIL **Design_** Straight Square Design   **Photography_** Li Guo Min   **Region_** Taiwan   **Client_** ARTIFACTS.TW.Co.,Ltd

WORK*shop* _ISSUE.TWO

Nine rotating glass doors at the entrance turn 360 degrees in all directions have given the flexibility to the display window.

ARTIFACTS DUNHUA BOUTIQUE

_Straight Square Design

Three large cubes are structured primarily crossing the store in order to promote the shop in shop concept, including the cubical display window floating in between indoor and outdoor. A specially structured LED installation in the display window and the logo are designed as an installation art for the entrance.

Floating effect has also been incorporated into the furniture. The islands are supported with transparent acrylic cylinders with white light spurting out from underneath; the shelves and clothing racks are half embedded in the wall with a distance from the floor; and most importantly the cubical display window is structured with an elevation from the ground and neon lighting effect underneath.

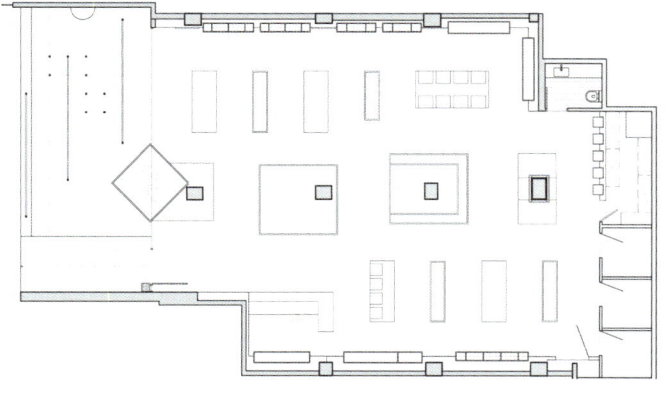

_PLAN

Design_ Straight Square Design **Photography_** Li Guo Min **Region_** Taiwan **Client_** ARTIFACTS.TW.Co.,Ltd

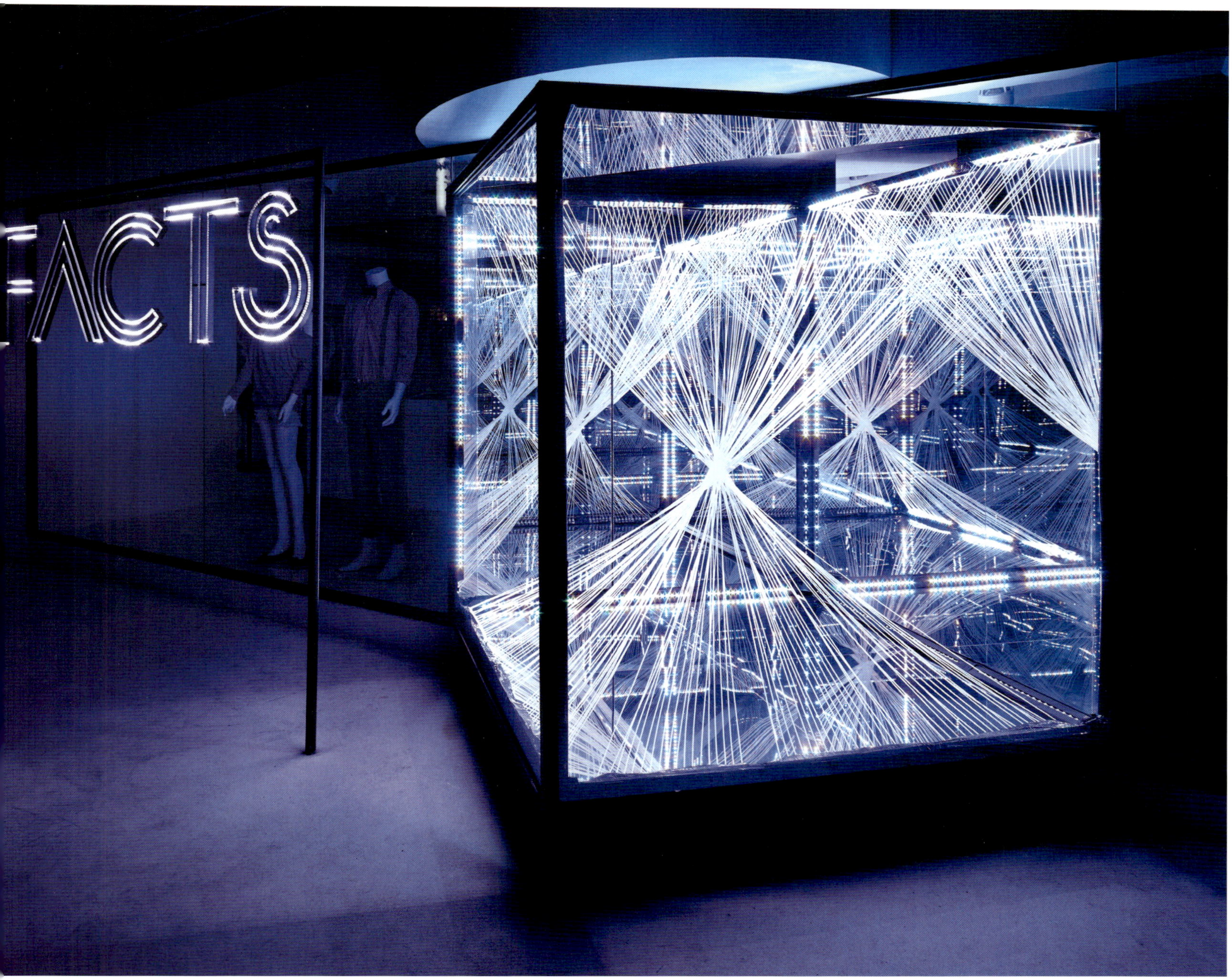

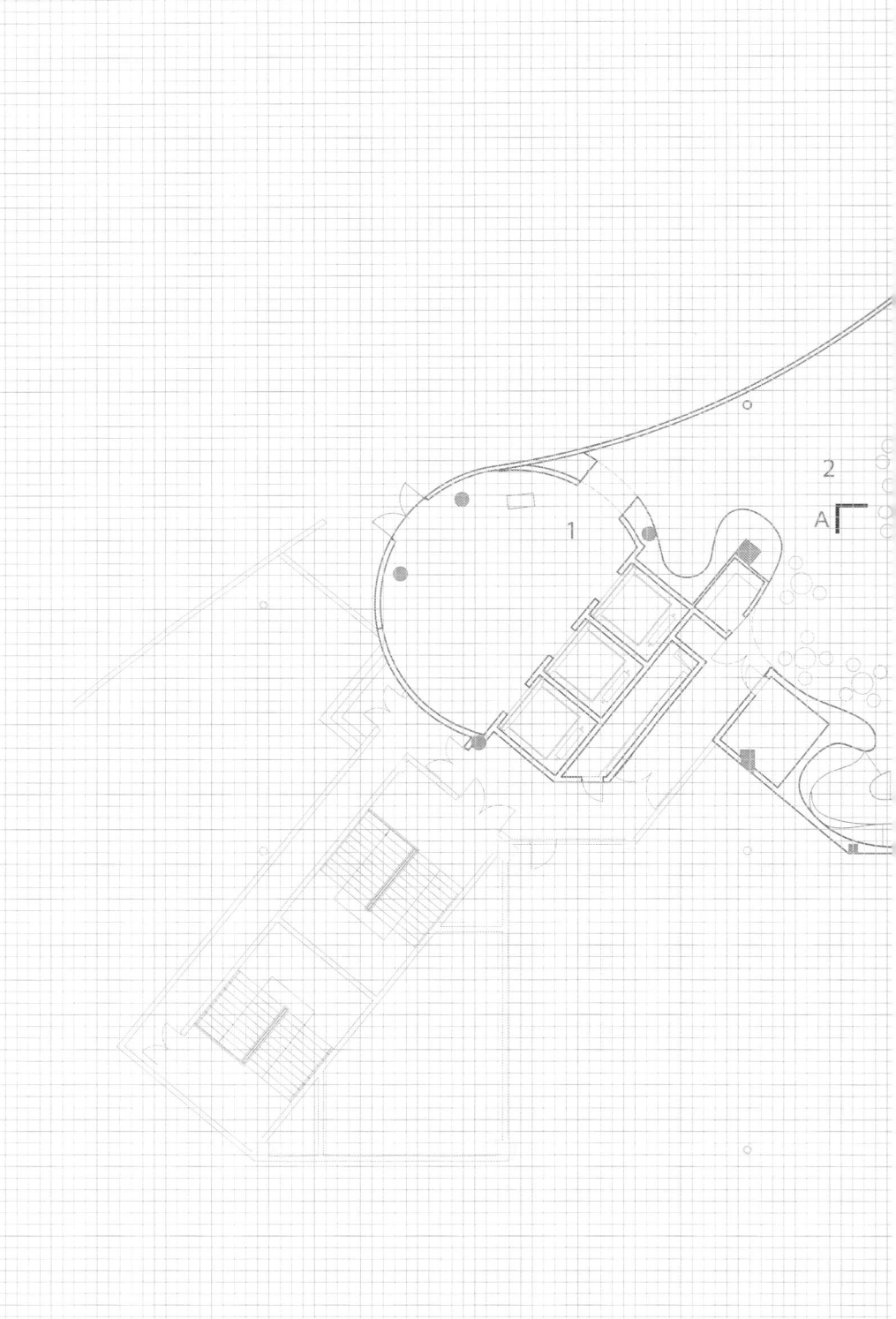

02
RESTAURANT/BAR

_GENERAL SOUTH VIEW

_GENERAL NORTH VIEW

_BAR

CIENNA ULTRA LOUNGE

_bluarch architecture + interiors + lighting

Cienna Ultra Lounge is a venue in the heart of Astoria (Queens, NY) designed to offer the ultimate entertainment experience to the discerning New York crowd.

If Cienna Restaurant is an exercise on experiential lightness, Cienna Ultra Lounge is an essay on softness. This project is conceptually linked to the mathematical clarification of the shape of a silk cocoon via a cosine Fourier series. Formally, the space transposes such geometric formalization into a meandering system of openings on the ceiling and walls.

The space is a cocoon entirely made of a seamless, tufted upholstered shell. The distribution of the 8,888 buttons defining the tufting was organized via 3D software, and allowed for precise fabrication and permissible tolerances. The system of openings in the ceiling and walls of this shell is punctuated by 88,888 acrylic strands.

_CEILING

Behind them is a system of LED light fixtures softly flooding the strands and the space with warm hues... and responding to music impulses via ad-hoc software. The space has reachable boundaries that offer a soft, tactile, sexy experience... The seating is made of sensuously outlined booths continuing the tufted shell. The tables are custom made in a sumptuous, full profile of poplar wood, and are lacquered in a gloss finish. The 45-foot bar shares its design with the bar at Cienna Restaurant. It is a lighted prism of honey onyx, and balances the DJ booth across the venue. The DJ booth also functions as a stage for live performances, as its sides are removable. The sound system is state-of-the-art and is fully integrated in the tufted shell. The sound system, the lighting system, and all other design systems are experientially and technologically inter-connected, as they exist and support one another to offer a seamless overall performance.

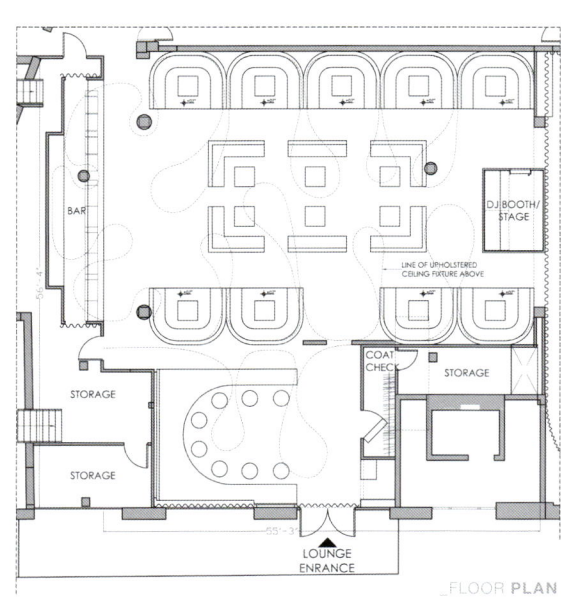

_FLOOR PLAN

TREE
RESTAURANT

_Koichi Takad Architects

We propose a dining concept that recreates Hanami, the traditional Japanese festival of the cherry blossom in bloom. Dining under the cherry blossom trees is a social gathering that celebrates the arrival of spring. This concept not only represents the serving of Japanese cuisines, but also hopes to capture a symbolic place for the locals to gather and dine under 'one big tree' and for the owner to nurture the business as if growing a tree.

We wish to emulate the comfort and tranquility the canopy of tree can create. Timber profiles create the branches of the tree, transforming the sushi train restaurant into a place of nature. Dappled light filters between the timber branches.

The flairs of light change as you move throughout the restaurant, mimicking the irregularity of natural sunlight, while highlighting the path of the sushi train.

Conceptually the Tree has become symbolic of the nurture and care put into growing this successful business. The branches extend to the perimeter, encompassing diners and workers alike. The timber profiles have been cut using CNC technology, minimizing waste and allowing accuracy and detail in the design. Gaboon marine plywood, brings the warmth of timber to the interior, which complements the texture of the rendered walls. The contrast of these elements highlights the central Tree and the sushi train below.

RESTAURANT / BAR   **Design_** Koichi Takada Architects   **Photography_** Sharrin Rees   **Country_** Australia   **Client_** Mr. Goro Usui

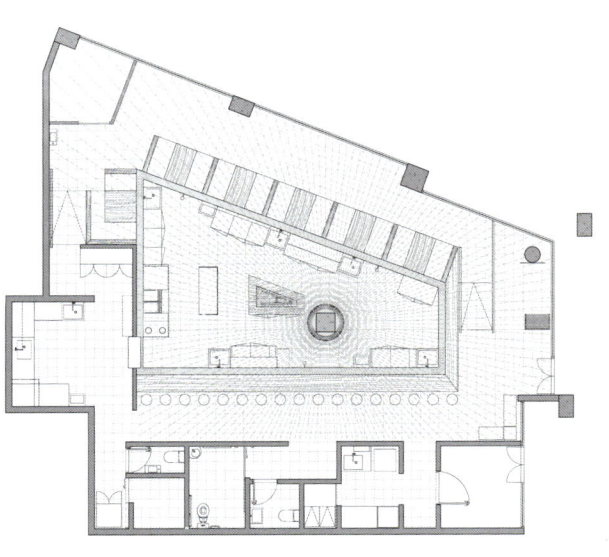

_PLAN

TANG PALACE, HANGZHOU

_Aterlier FCJZ

The restaurant is located on the top floor of a superstore in the new town area of Hangzhou, with 9m storey height and a broad view to the south. Composite bamboo boards are selected as the main material, conveying the design theme of combining tradition and modernity.

In the hall, to take advantage of the storey height, some of the private rooms are suspended from the roof and creating an interactive atmosphere between the upper and lower levels, thus enriching the visual enjoyments. The original building condition has a core column and several semi-oval blocks which essentially disorganised the space. Hence, our design wants to reshape the space with a large hollowed-out ceiling which is made from interweaved thin bamboo boards; and extending from the wall to the ceiling, the waved ceiling creates a dramatic visual expression within the hall. The hollowed-out bamboo net maintains the original storey height and thereby creates an interactive relation between the levels. We also wrapped the core column with light-transmitting bamboo boards to form a light-box, which transforms the previously heavy concrete block into a light and lively focus object.

The entrance hall also follows the theme of bamboo. The wall is covered with bamboo material which follows the original outline of the wall, turning it into a wavy surface. In this way, the surface echoes the hall ceiling as well as performs a guiding function for customers.

The design of private rooms embraces different characteristics. The rooms on the first level are relatively bigger and share the features of expanded bamboo net from the wall to ceiling and ornamentally engraved wall surfaces. Meanwhile, the different folding angles and engraved patterns make each room different from one another. The rooms above on the south are smaller and feature a special waved ceiling pattern and simple bamboo wall surface, which creates interesting and spacious room features. The key design concept of the space is that the suspended rooms are connected with suspended bridges and sideway aisles. The semi-transparent wall provides a subtle relationship between the inner and outer spaces, bestowing people with a special spatial experience.

In this design, we hope to create diversified and yet an interrelated interior spaces through the different usages of the new bamboo material, responding to the local culture while seeking intriguing spatial effects.

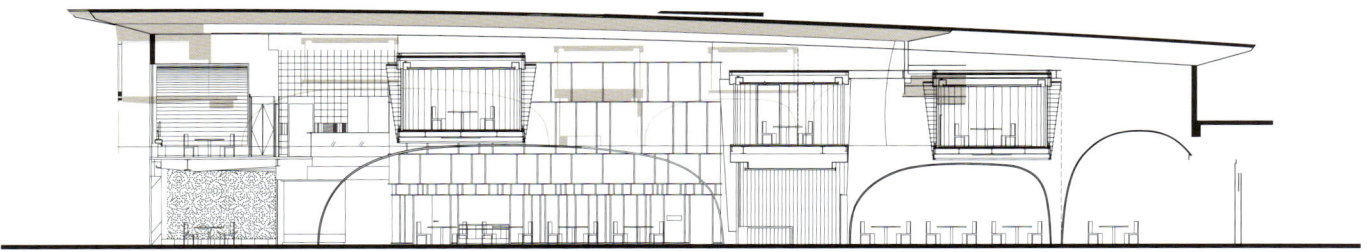

_SECTION

Design_ Aterlier FCJZ **Photography_** Shu He **Country_** China **Client_** HongKong Tang Palace Food&Beverage Group Co., LTD.

EVERYDAY CHAA

_MAEZM + Sarah Kim

Everyday Chaa is a new concept of franchise commercial space with the theme of Korea's 'traditional tea'. What client wanted between the saturated cafés of Seoul was a space where consumption was made in more convenient, friendlier, and more modern way through traditional drink rather than coffee. Therefore, the space didn't have to be traditional, and we hope it to appear in Seoul quietly but strongly where cafés are lining. The space is in gradation from the bottom to the ceiling. The gradation rising from the bottom to the wall crumbles all factors within the space vaguely. As if it is standing in the middle of desert or a space casted with deep choreography. We wanted the space itself to show depth at a clean structure that did not give change in forms and a space without any special details. We hope that this would be sufficient suggestion to an 'act of drinking tea' as imagining a space where a light turns up like moonlight casts in the middle of silence and where leaves are in full glory.

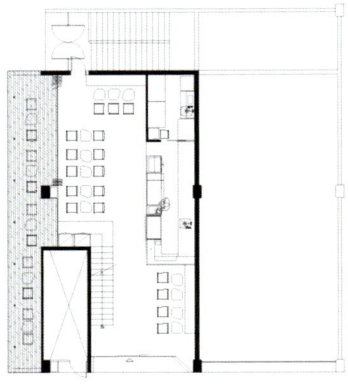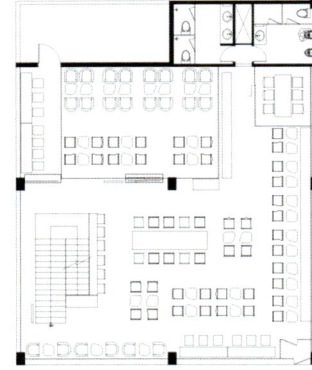

_PLAN

RESTAURANT / BAR   **Design_** MAEZM + Sarah Kim   **Photography_** Shin Tai Ho   **Country_** Korea   **Client_** iCare inc.

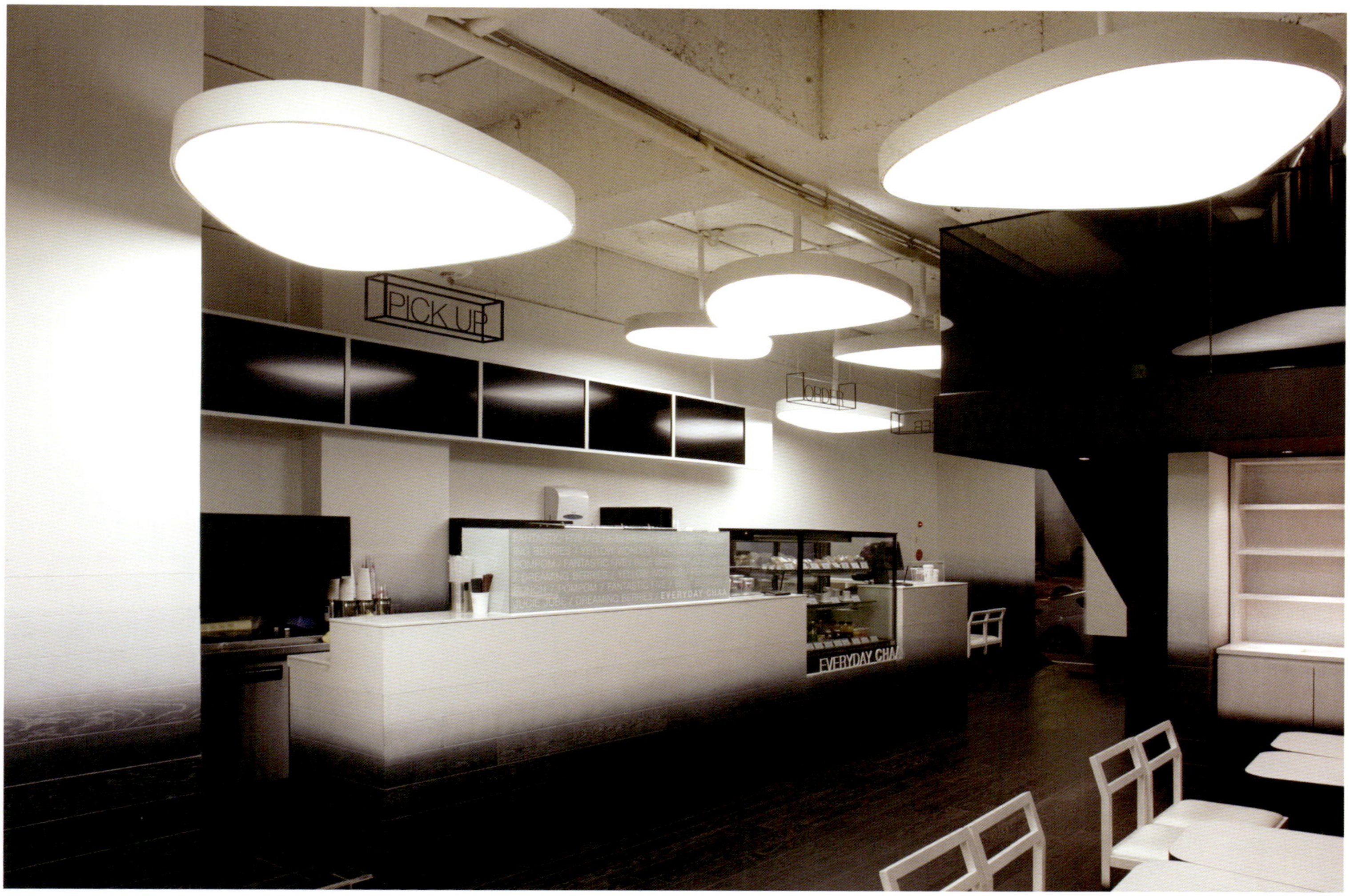
PICK UP
ORDER
EVERYDAY CHAA

HERZEN
EVERYDAY CHAA

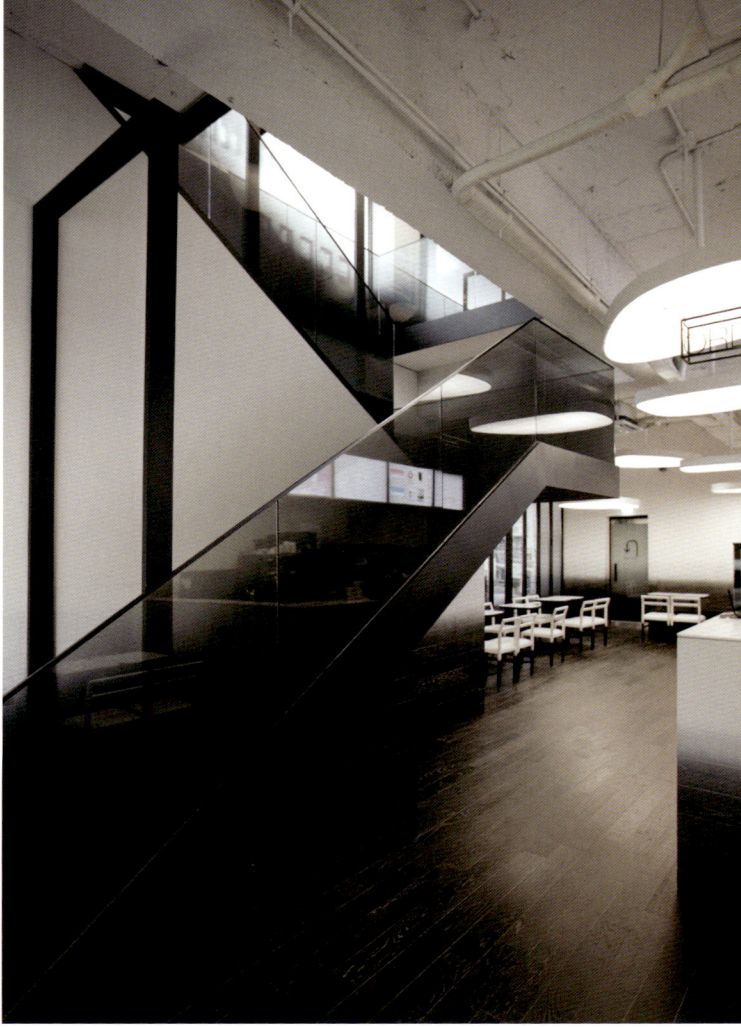

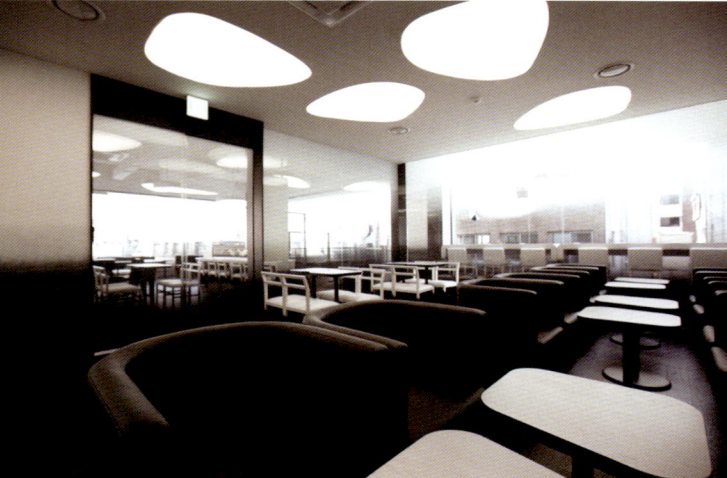

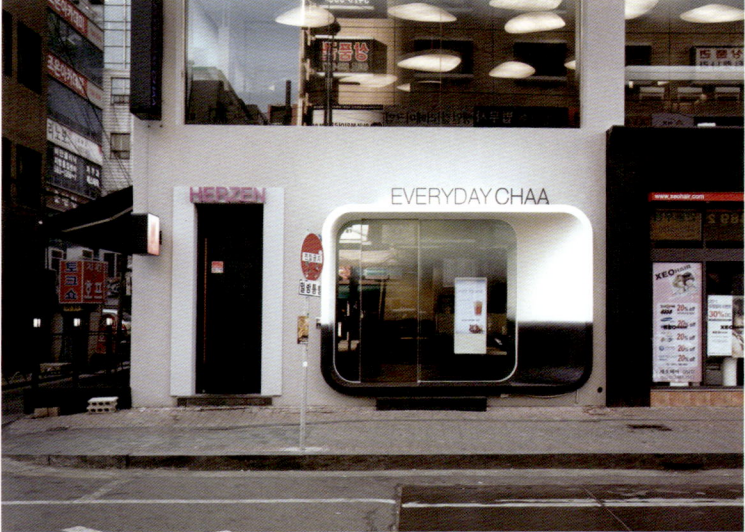

WORK*shop*

_ISSUE.TWO

NISEKO VILLAGE LOOK OUT CAFÉ

_design spirits co., ltd.

Somewhere near to top of the mountain in Niseko on the island of Hokkaido in Japan, there is the Look out café for skiers to have a short break and lunch. The restaurant opens only during the snow season – December through April daily. It was originally built of wood since 28 years ago, and now this is the first time of its renovation.

I was informed about the restaurant renovation in Niseko On 4th October, 2010. Initially, the restaurant owner hoped its opening in early December. After a few meeting going on and also conference call to the project manager in Hokkaido, I was requested to complete the project before mid-November due to snowfall begins during that period. Counting from the day of the meeting to mid-November, it was approximately 45 days left.

It took 4 days to determine the overall design and materials, and got approval from the project manager on the fifth day. At the same time, we prepared drawings and visited the site in Hokkaido. After some comparisons and discussion going on, we have selected a suitable contractor to work together. Thereafter, constructions progress is immediately proceed.

Since the Look out Café is located on the top of the mountain, it is impossible to reach it by a car or ski-lifts during the off-season. There are many times that the constructors and me were walking, hiking, and climbing over another slope to get down from the site. In the meanwhile, materials were carried by caterpillar vehicle with a carrier attached to it, the other construction workers had to climb up and down the mountain on foot if the carrier is fully-occupied with materials. Also there was situation when it rains we were prohibited to climb the mountain as there is possibility of land sliding.

In Japan, daytime is shorter than nighttime in autumn and winter whereas sunset begins at 16:00 in despite of sunny day or cloudy day. In addition, strong winds also come as a result of frigid condition, which forced us to leave the site before sunset to avoid any unfortunate consequence. As a result, construction progress unable to proceed as scheduled. Fortunately, it was a warm winter as snowfall came late, and construction could be carried out until late November. The project is completed by early December, with the grand opening of Look out Café.

Although there are many constraints during the progress, but the outcome is very satisfying and impressive. Not to mention about the tight deadline and managing the construction workers, limitation in using only three types of materials – woods, paint and wallpaper is also a challenge.

We used vertical timber lattice as the main material, which is known to represent Japanese identity. So, now, the tourists can feel the exotic Japan when they are walking along the alleyway, as the reflections from the roofs come in different sizes and heights.

Illumination effect is created as the lights from roof came through the lattice. And the feeling of warmth and secure are all around the space as it is surrounded by roofs with various sizes and heights. The feeling is more obvious especially when the space is crowded. It was attempt to create an ambience whereas can feel the outside world atmosphere in an interior space.

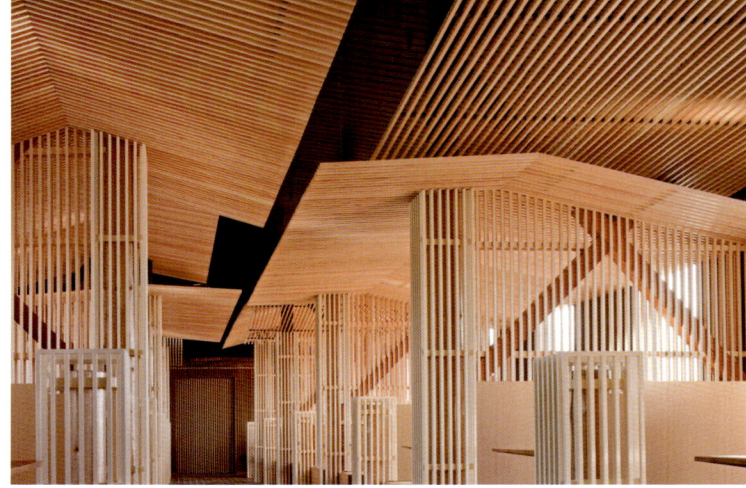

RESTAURANT / BAR   **Design**_ design spirits co.,ltd.   **Photography**_ Toshihide Kajiwara   **Country**_ Japan   **Client**_ YTL Hotels

_PLAN

HAPPO-EN THE HAKUHO-KAN

_hashimoto yukio design studio inc.

Happo-en is a wedding facility with an expansive Japanese garden. The Hakuhou-kan is the name of the authentic Japanese-style architecture that has been sitting through the ages inside this graceful garden. The concept we proposed when renovating this building was to fuse the traditional with the future. Our aim was to harness traditional architectural design while introducing unconfined, unprecedented space design. Another aim was to project the image of the Hakuhou inside the entire space, the white phoenix for which the building name 'Hakuhou-kan' is named after.
Squared timber was used on the ceiling of the banquet area to create a dense structure, which alone, expresses the widespread wings of a white phoenix and brings to life a delicate yet dynamic space. The carpet represents a sea of clouds that calls forth images of the white phoenix soaring high in the air. On the front wall facing the entrance and on its opposite wall where the stage stands is kimono fabric adorned with a pattern of a white phoenix, created by Kyoto kimono designer Jotaro Saito.

MAIN ROOM

FOYER

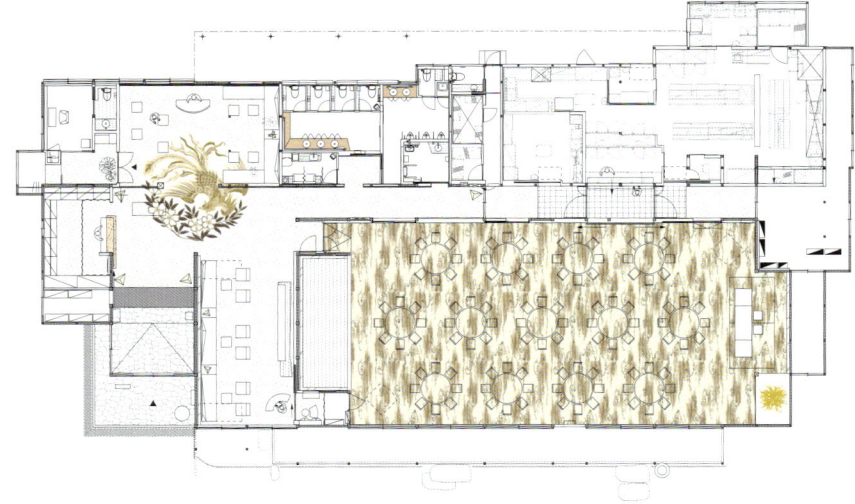
PLAN

The ceiling, walls, and furniture inside the foyer was created with the concept of origami that brings both stateliness and airiness to the space. The form is airy like origami but the materials used within such as kimono fabric and gold leaf bring a certain depth to the entire expression.

As such, materials traditional to Japan and mythical images that have been passed down for generations in this culture have been fused with a free form of expression to create a new Japanese-style space that looks beyond into the future.

ENTRANCE

TWENTY FIVE LUSK

_CCS Architecture

On Lusk Alley in San Francisco's South of Market district, a 1917 smokehouse and meat-processing facility has been renovated to become Twenty Five Lusk. The 265-seat new American restaurant and bar is an unexpected gem in the urban fabric. CCS Architecture crafted the two-level space, weaving graceful forms and sophisticated materials through the massive, historic, warehouse structure. The interior emphasises a counterpoint between the new palette of polished stainless steel, glass, white plaster, leather, mirror, faux fur, and slate and the existing elements of brick, concrete and rough-sawn timber.
The architects created a large, glass entrance, cutting into the existing building exterior; the canopy bends up at its leading edge to become the restaurant's signage. Windows were enlarged and added along the façade to animate the interior with natural light and allow views. Inside, a large wedge from the upper floor makes an open connection between the lower level lounge and the dining room upstairs. Entering the restaurant, guests take in simultaneous views of both.

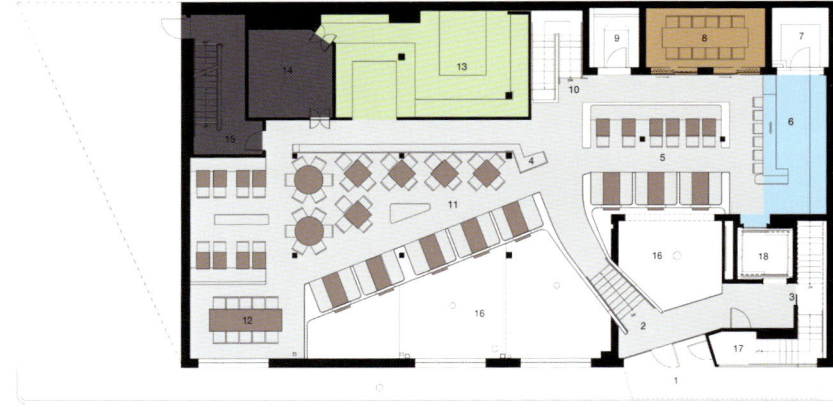

_FIRST FLOOR **PLAN**

1_ MAN ENTRY
2_ STAIRS TO MAIN DINING
3_ BACK STAIRS TO LOUNGE
4_ HOST
5_ BAR SEATING
6_ BAR
7_ LIQUOR STAORAGE
8_ PRIVATE DINING
9_ WINE ROOM
10_ MAIN STAIRS TO LOUNGE
11_ MAIN DINING
12_ COMMUNAL TABLE
13_ GLASS- ENCLOSED KITCHEN
14_ WARE WASH
15_ EXIT STAIRS
16_ OPEN TO LOUNGE BELOW
17_ STAIRS TO OFFICE
18_ ELEVATOR

MAIN DINING + ENTRY
BACK OF HOUSE
KITCHEN
BAR
PRIVATE DINING

RESTAURANT / **BAR** **Design_** CCS Architecture **Photography_** Paul Dyer **Country_** USA **Client_** Matthew Dolan, Chad Bourdan and Chris Dolan

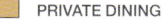

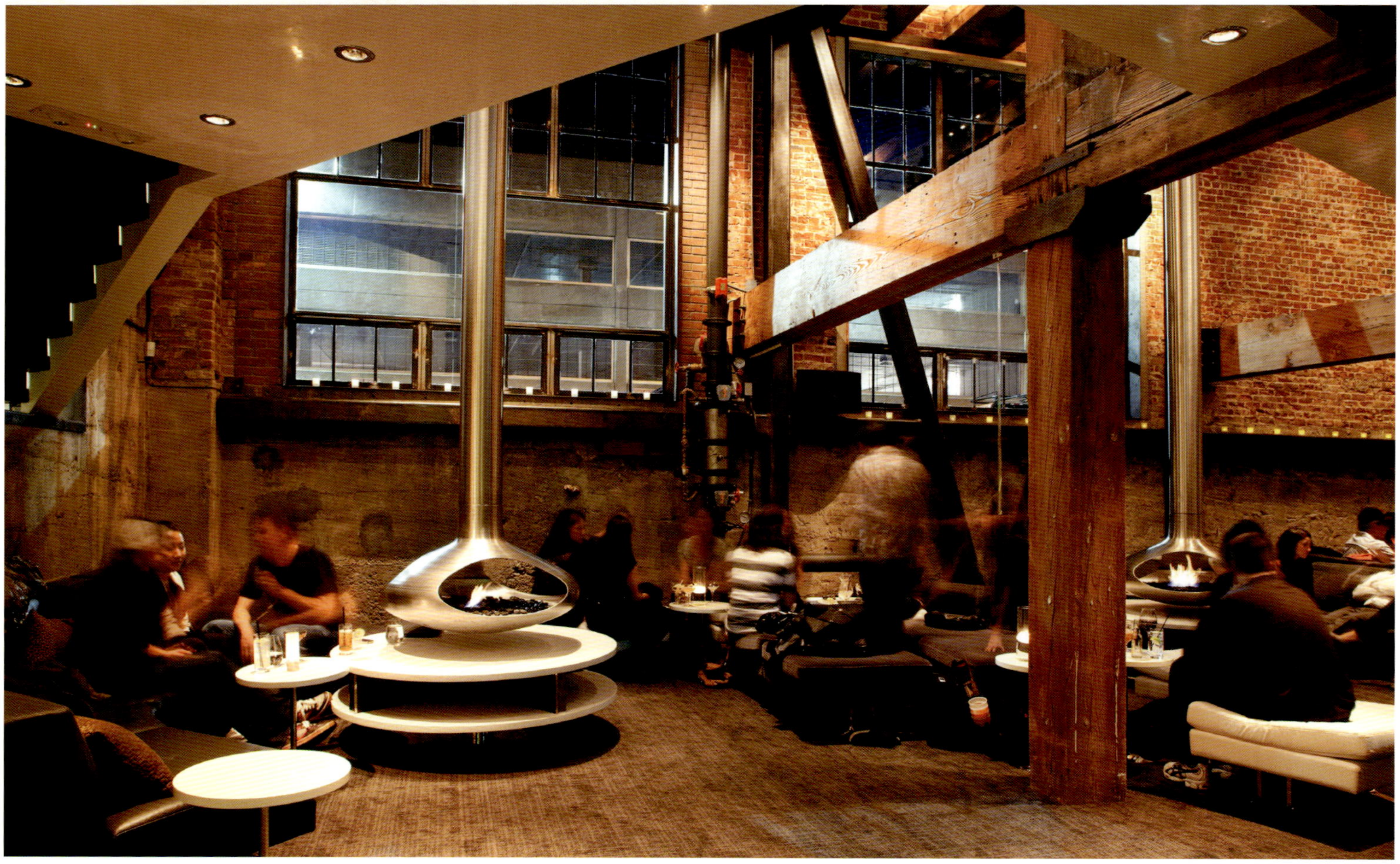

The dining room is on the second floor, up a half-flight of stairs from the entry. The kitchen is a highlight on this level: a modern envelope of clear and black glass permits views of the chef action and reflects the activity of the dining room. A strategic mix of tables, banquettes, and booths provides seating for 120. Pullman-style booths are built into the angled, low plaster wall that borders the cut-away, and cantilevered tables, made from richly patterned Macassar ebony, pierce the wall. Lighting reveals the original Douglas fir ceiling and creates a warm glow.

In the lower level lounge, seating zones extend the length of the space, each with a suspended, stainless steel fire orb. The orbs act as a focal point for each seating area, much like camp fires, and their reflective flues extend up through the restaurant's open spaces to the ceiling 20 feet above. Behind the large bar, the former Ogden Packing and Provision smoking rooms have been converted into intimate lounge areas. These semi-private, brick and concrete chambers are appointed with sumptuous sofas. The lower level features a 40-seat private dining room as well as a glass-enclosed wine room within the former freight elevator shaft. The architecture sets up a notable contrast between the dramatic vertical space and the single-height areas, allowing guests to experience the restaurant in its totality while providing intimate spaces to explore. CCS transformed the entire 15,000sqft warehouse to accommodate new uses. Twenty Five Lusk occupies the first and second floors, and the third level has been designed as 5,200sqft of creative office space.

SANTA RITA RESTAURANT

_Pedro Pacheco

The intervention site is located in Lisbon's historical centre, on the ground floor of a Pombalinian building. It is a very common type of warehouse in these buildings, characterised by a single space defined by three vaulted areas with stone masonry walls and solid brick domes. Its maximum height is 4.15m and it is 16m long, 5.50m wide, having a total area of 102m². One of the ceiling tops has direct contact with the street, via two doorways, while the other has contact with a ventilation patio.
The architecture project is meant to remodel the existing space for it to become a restaurant, by keeping and outlining its space and architectonic qualities: to maintain the longitudinal reading of the vaulted space, characterised by the lioz limestone arches and pilasters.

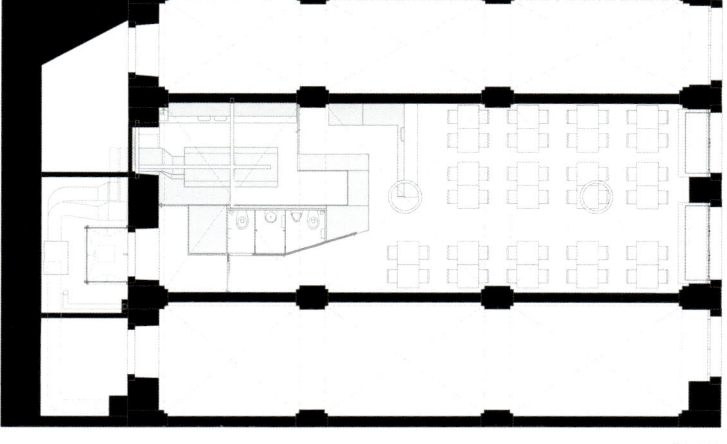

_PLAN

RESTAURANT/BAR **Design**_ Pedro Pacheco **Photography**_ FG+SG Fotografia **Country**_ Portugal **Client**_ Luís Henriques e Conceição Henriques

The project options are then born from the need to create an element capable of both separating and connecting the dining room and the kitchen, visually freeing the arches and vaults throughout the space's total length and introducing in the room light and reflection.
This screen-element is defined by a system of white-lacquered glass panels, aligned with the stone arches' offset, that circumvent the kitchen's different work areas and the customer washrooms, restraining access to the service area and defining the restaurant's new image and environment. The street light becomes twice part of the room atmosphere, as the street's moving images and reflections are being projected onto the screen. Simultaneously, indirect and suspended LED luminary floods the room with colour and regulates its temperature.

WORK*shop* _ISSUE.TWO

M.N.ROY CLUB

_Emmanuel Picault, Ludwig Godefroy

M.N.Roy is a project made as an open question, the one has for goal not to answer obviously what's actually the M.N.Roy.
In this way, the place can be perceived as an anti-project of what could be the commission of a private club in Mexico City, and more precisely in its Roma neighborhood.
In fact Roma has been very important in the definition of the architectural identity of this space, located in a very dueling neighborhood, and responding on one hand to its past, the one of the high mexican bourgeoisie of Porfirio Diaz (Mexican dictator 1876-1911) time which abandoned the neighborhood after the 1985 earthquake, and today's reality of a trendy urban area that Roma became. The club is the expression of a high singular personality settling in the strong left-over of its past time.
According to this, where normally the renovation of the façade appears to be the starting point, the opposite was done: letting the façade untouched to increase the rupture between the original meaning of the house and the redefinition of it.

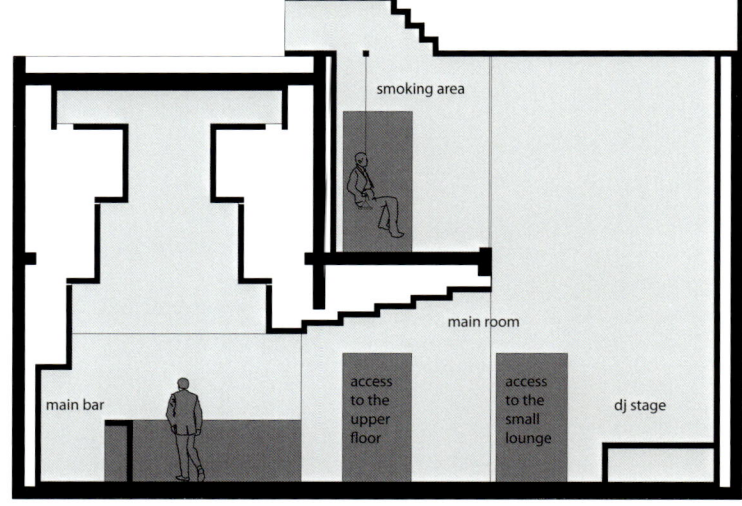
_SECTION

RESTAURANT/BAR   **Design**_ Emmanuel Picault, Ludwig Godefroy   **Photography**_ Ramiro Chaves   **Country**_ Mexico   **Client**_ Private

WORK*shop* _ISSUE.TWO

We kept the house as a testimony of what it was, the house where Manabendra Nath Roy founded the first clandestine Mexican communist party.
By not touching the façade we made paradoxically appealing the building from outside, stimulating the curiosity of the people passing by and seeing a large queue trying to enter an almost ruined house.
Once inside, we made another step in a schizophrenic architectural way, introducing a new language, deeply belonging to the mexican culture, and nevertheless completely stranger to the Porfirio Diaz architecture time.
We used a pre-hispanic language reminiscence inside, in a participative way and not contemplative as could be a nostalgic neo pre-hispanic vision of it, introducing new materials (copper, leather, wood, volcanic stone), geometries (puuc art, maya arch, pyramids), and everything, down impressive generous volumes.
M.N.Roy is the impossible mix of cultures, volumes, architectural styles, making possible an improbable modern space of melting pot.

WORKshop

_ISSUE.TWO

ZEBAR
A LIVE BAR
IN SHANGHAI

_3GATTI

This project was born in 2006 when a Singaporean movie director and an ex musician from south of China decided to open a live bar in Shanghai. The budget was very low but the client was incredibly good and open-minded to us.
The schedule was very tight and fortunately they liked immediately one of the first concepts I proposed to them: a caved space formed from of a digital Boolean subtraction of hundreds of slices from an amorphic blob. The idea looks complex but actually is very simple and was born naturally from the digital 3D modelling environments where me and others enjoy playing with virtual volumes and spaces. The space was subdivided into slices to bring it back from the digital into the real world, to give a real shape to each of the infinite sections of the fluid rhino nurbs surfaces.

In Europe the natural consequence of this kind of design will be giving the digital model to the factory and thanks to the numeric control machines cut easily the huge amount of sections all different from each other.
But we were in China where the work of machines is replaced by the work of low paid humans. Using a projector they placed all the sections we drew on the plasterboards and then cut each of them by hands. The cost was surprisingly low and the sense of guilt towards the workers higher.
The construction was incredibly fast and was almost finished in a couple of months, when we discovered the naïve clients didn't have any business plan and the site remained closed for 3 years and was finally completed and opened in 2010 when they discovered how to run the business.
This is the story of Zebar, a digital design built into an analogic world.

RESTAURANT / BAR   **Design_** 3GATTI   **Photography_** Daniele Mattioli   **Country_** China   **Client_** Jim Dandy

_ZEBAR **A LIVE BAR IN SHANGHAI** _RESTAURANT/BAR

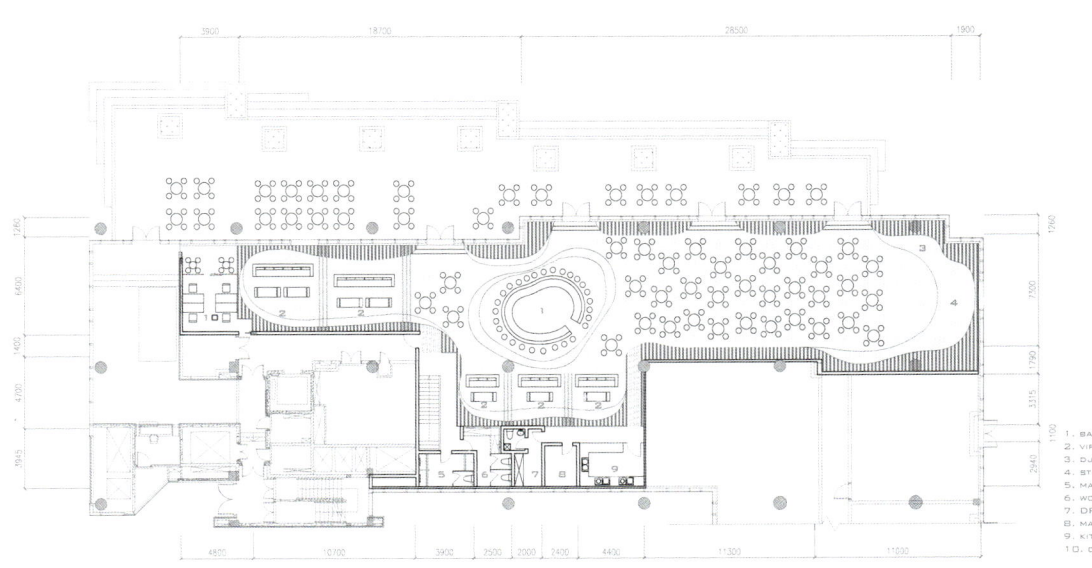

_PLAN

PHILL

_Nuca Studio

Phill is a meeting place designed for the entire family. It has a playground, a multipurpose room and a small café at the ground floor and a gourmet restaurant at the first level.
The playground and the multipurpose room are enclosed areas with independent light and acoustic scenarios and they accommodate activities from theatre and puppet shows to martial arts and ballet lessons. Upstairs, the dining area is an open space directly linked with the lobby. In between them the small café communicates visually with the playground through a couple of round openings. The functions of the program have their own agenda but at the same time they work closely together therefore the connection of the individual spaces was very important. In order to link these different rooms, the walls were perforated by transparent openings and a special attention was paid to the design of the stairs which climb their way to the first floor around a four meter tall elephant.

_RECEPTION

RESTAURANT/BAR — Design_ Nuca Studio — Photography_ Cosmin Dragomir — Country_ Romania — Client_ Phill

WORK*shop* _ISSUE.TWO

_DINING **AREA**

_DINING **AREA**

_CAFÉ ON THE **GROUND FLOOR**

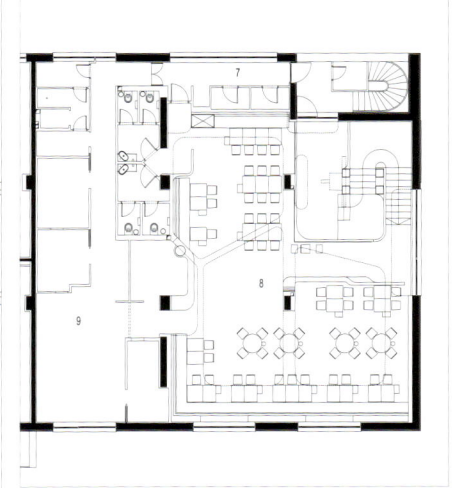
_FLOOR **PLAN**

1_ MAIN ACCES
2_ LOBBY
3_ RECEPTION
4_ PLAYGROUND
5_ CAFÉ
6_ MULTIPURPOSE ROOM
7_ TECHNICAL & STORAGE
8_ DINING AREA
9_ KITCHEN
10_ BACKYARD

We went out subjectively, choosing references that were appealing to us, and then adapted the concept for the space in study, hoping other people would share the same perspective. The premises for picking the design typology was, alongside our personal passion for comic books, animation and vinyl toy subcultures, the idea that kids' perception is anything but reductionist and adults are not that sober not to engage in any imaginative exercise. The elephant's role and the overall geometry are for interpretation, even though we have had our personal story to tell, it is far more interesting to observe the visitors creating their own response.

The space's layout was mostly determined by: dividing two groups of activities specifically for both kids and adults, binding the two, distributing the space on the two levels and the lavish backyard.

WORK*shop* _ISSUE.TWO

WIENERWALD

_Ippolito Fleitz Group

Friedrich Jahn opened the very first Wienerwald restaurant in Munich in 1955. The synonymous fast-food chain expanded over the following decades until it was operating branches in 18 countries.
Following the collapse of the group, the company was under varying ownership until the grandchildren of the founding family bought back the rights to the brand in 2007. Their goal is now to build on the long tradition of the company, exploiting both the strength of the brand and the uniqueness of their gastronomic concept. Our studio was commissioned to develop new corporate architecture for the chain, which has already been rolled out in two Wienerwald branches in Munich.
The new interior design underscores the realignment of the brand, while translating the chain's traditional strengths of high quality, comfort and German cuisine into a contemporary design idiom. Materials and colours reflect the principles of freshness and naturalness, which find their expression in materials such as wood, leather and textiles, as well as in the dominant green tones that complement the fresh white. The space has been organised to ensure good visitor guidance, crucial in a self-service restaurant, as well as respecting the need for a differentiated selection of seating. Upon entering the restaurant, the guest is guided towards a frontally positioned counter, which presents itself as a clearly structured, monolithic unit.
In front of the service counter is a service station made of white solid surface, offering sauces, condiments and cutlery. It stands on golden chicken legs and looks expectantly towards the entrance. Green instructions and Wienerwald chickens set into the rustic wood floor

RESTAURANT/BAR **Design_** Ippolito Fleitz Group **Photography_** Zooey Braun **Country_** Germany **Client_** Wienerwald Franchise GmbH

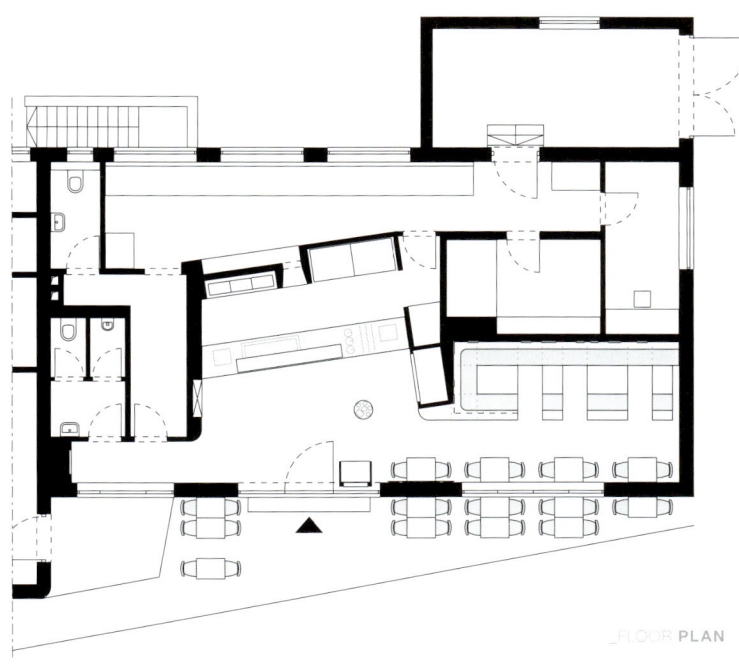

_FLOOR PLAN

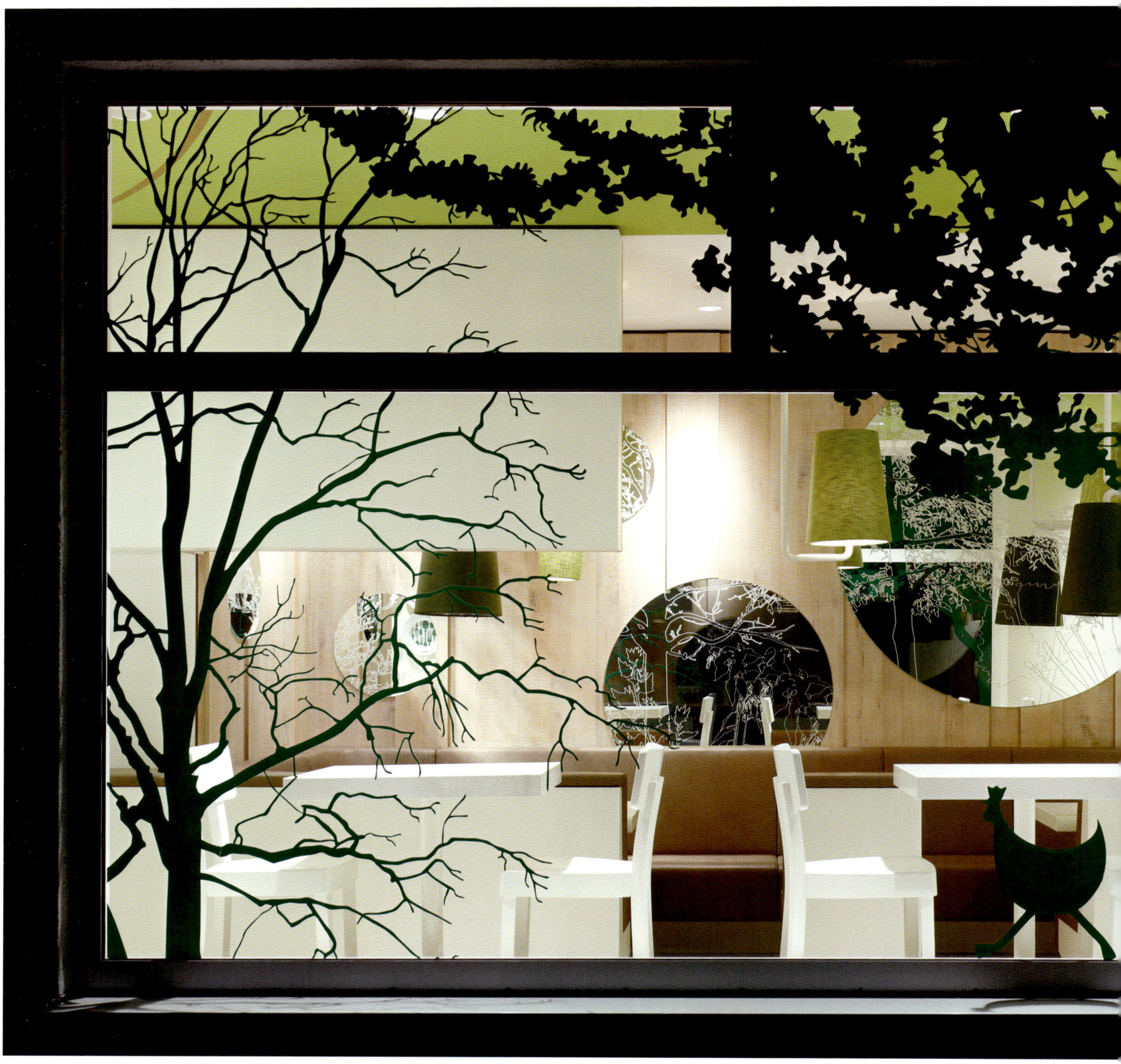

Heute bleibt die
Küche kalt,
wir gehen in den
Wienerwald.
Wienerwald
Wienerwald
Wienerwald

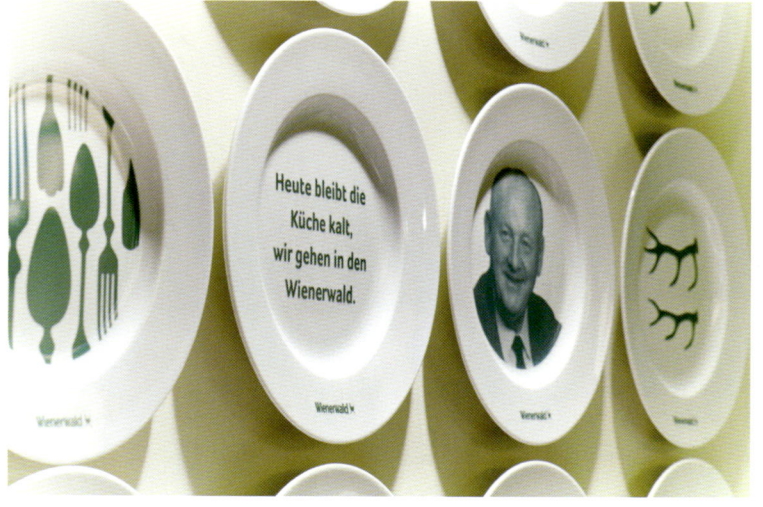

show the customer how to navigate the ordering process.
The dining area offers a range of seating options catering toward different requirements. White solid surface high bar tables are available for guests with little time on their hands. These are supported by a single leg with a tapering cylinder at its foot, recalling the traditional turned table leg. Alternative seating is available in an elongated seating group upholstered in brown, artificial leather, a reflection of the traditional Wienerwald seating niches. Guests are really spirited away into the 'Wienerwald' (English: Vienna Woods) here. Overlapping, rough-sawn oak panels on the rear wall quote the forest theme. Round mirrors printed with the outlines of tree and forest motifs are set into this wall. Different-sized pendant luminaires at varying heights hang over the tables. These are sheathed in a roughly woven fabric in three shades of green and ensure a pleasant atmosphere.

Forest images in different shades of green on wallpaper occupy one side wall, as well as transparencies on the windows.
The view into the restaurant from the outside thus becomes a multi-faceted experience in which the individual elements on the mirror and glass surfaces reflect and overlap one another, making the brand world a truly holistic experience.
A display of dining plates on the wall is dedicated to the Wienerwald company and its long tradition, reminiscing on the history of the brand in 14 motifs. They pay tribute to Friedrich Jahn, the brand's founding father, and show a photograph of the first Wienerwald restaurant.
The new restaurant design repositions Wienerwald as a contemporary fast-food chain. Traditional elements of the brand have been incorporated and translated into modern spatial elements with an exciting twist.

HOLYFIELDS FRANKFURT

_Ippolito Fleitz Group

Holyfields, a wholly new restaurant chain concept, commissioned our studio to develop a modular, scalable space system with a distinctive look and feel. The brand's claim 'time to eat' describes an innovative concept based on a sophisticated ordering system that gives diners more time to eat. The restaurant guest orders at one of ten touch screens in the entrance area, which show the menu in image and video format. He then takes an electronic signaller with him to his seat. This emits a signal when the food is ready to be collected from a central counter.

The dining room contains a wide variety of seating which is staged in four tiers that are staggered in height from the front windows to the rear wall. The next tier is created by a row of white tables with upholstered, two-seat benches. These five table groups are further demarcated by the slightly raised, dark-wood plinth and the gently lowered ceiling above. The next tier offers guests a seat at a long, bleached oak bar table, contained between columns, in the very busiest area of the restaurant. Finally, four white, six-seat tables at the same height as the long bar table are aligned with the rear wall, which is executed in dark wood slats. Capacious U-shaped enclosures give a final parenthesis to the space. This area affords the best view across the entire room from a slightly more retired position. The open-plan space means the visitor can see the far end of the longitudinal axis from the entrance area. This far wall is home to the food counter, prominently encompassed by a funnel-shaped, floor-to-ceiling copper wall. The food counter is more like the buffet at a party than a traditional serving counter. Here food distribution is celebrated in style. Kitchen hatches and glass rear walls give the guest a glimpse of the busy bustle of the kitchen. The prominence of the food counter is further enhanced by three illuminated ceiling elements like airport signs that give names to the three serving counters below: Peter, Paul and Mary. The guest receives the name of the respective counter on his electronic signaller and so knows exactly where to go to pick up his order.

_THREE NAMED SERVING **COUNTERS**

RESTAURANT/BAR **Design**_ Ippolito Fleitz Group **Photography**_ Zooey Braun **Country**_ Germany **Client**_ Holyfields Restaurant GmbH & Co. KG

A fountain of white terrazzo stands in the entrance foyer. Here guests can help themselves to a glass of water free of charge. Drinks and desserts can also be ordered separately from your main food order at the bar which is crafted from dark-stained oak with a black leather-clad front. The bar opens onto the dining room, but also caters to a smaller lounge located on the other side of the host counter.
The lounge consists of a modular system of armchairs and poufs in different warm leather tones, complemented by occasional tables with integrated textile lampshades.
The restaurant concept is complemented by a take-away area which is accessed via a separate entrance and is divided from the lounge by large glass shelves. Much attention was invested in the acoustics of the dining room. A specially commissioned acoustic ceiling with geometrically patterned holes guarantees good acoustics. It also creates an attractive visual counterpoint to the raw concrete, floral patterned floor that runs throughout the space, serving as one of the main key visuals of the new restaurant.
The first Holyfields branch in Frankfurt's Kaiserstraße will be followed by new openings in premium locations in other German cities such as Berlin, Stuttgart and Hamburg.

SMITH&HSU TEAHOUSE

_Carsten Jörgensen

smith&hsu is a contemporary tea brand based in Taiwan. Its premium loose teas, collected from around the world, are a testament to its deep passion for both Chinese and British tea culture. Beside its carefully assorted tea collection, smith&hsu offers a wide range of beautifully designed tea tools and homemade gourmet food.

smith&hsu's teahouse on Nan Jing East Road in Taipei is the 5th and latest addition to the brand. Envisioned by Swiss / Danish designer Carsten Jörgensen, the new teahouse has two floors seating 48 guests in the upper dining area and 10 guests in the spacious lower tea shop. It carries minimalistic tea tools exclusively created for smith&hsu and its outstanding teas. The wood and concrete interior feels authentic. The materials chosen for the store reflect the subtlety of a great tea and trigger the guests' aesthetic sensibility. In keeping with modernistic principles of visual clarity and simplicity, Carsten Jörgensen has created a wonderful framework for experiencing quality teas. The teahouse's ascetic yet warm charm has a calming effect even after one of those long and stressful days.

RESTAURANT / BAR **Design_** Carsten Jörgensen **Photography_** Alain Kuan **Region_** Taiwan **Client_** smith&hsu Co., Ltd.

_FIRST **FLOOR**

_WOODEN CUBIC SHELVES ON THE **FIRST FLOOR**

As an extension of the design for the previous smith&hsu teahouses, the key elements of the new store are 'soil' and 'wood'. The store's concrete surfaces display a subtle spectrum of grayish, bluish, yellowish and brownish colours. Concrete walls and floors add an earthy feel, whereas the wood gives the store a sense of organic warmth. All the materials smith&hsu has used for the teahouse feel refreshingly raw and uncluttered.

The cubic wooden tables, counters and shelves are simple and unpretentious. On the first floor, Y Chairs by Hans J. Wegner and on the second floor, Eames Plastic Side Chairs by Charles & Ray Eames complement each other and the cubic furniture well. Both are epitomes of the 'designer chair' and both are exceptionally beautiful.

The sensuousness expressed in the Eames chair, its elegance and comfort, seems to have made it a perfect match for smith&hsu. Moreover, the inclusion of these two iconic chairs is a sure sign of the brand's desire to bring only the best to its customers.

Bookshelves made of piles of wooden cubes run around the walls of the entire second floor, creating an open library for smith&hsu's guests. The books come from the customers themselves and from a few generous donors. The tea and the books, the concrete and the wood somehow all make sense together in this great looking new teahouse. smith&hsu has managed to combine asceticism with homeliness and the result is best described as something akin to wisdom.

_SECOND FLOOR

_SECTION

+GREEN

_Chikara Ohno / sinato

An organic restaurant located on Jiyu Street, which is a short walk from Komazawa Park, one of the biggest parks in Tokyo. It has three functions such as a takeout, an organic food shop and a restaurant. The most distinctive feature of the premises lies in its floor level, which is 1.61m below the ground level of entrance. And the height of the premises is 4.39m. To study the arrangement of three functions in such a unique space was the starting point of our design.

I arranged the restaurant on the half underground floor, the takeout place in front of the entrance at same level as the ground and the shop place diagonally away from the takeout place not to hide the restaurant from the entrance and to get sunlight on the restaurant floor from façade opening. The shop place is 0.56m above the takeout place in order to arrange the kitchen under it.

There are 3 floor levels for each function, so people move up and down in this space. I made different walls in upper space and lower space and partitioned space in different ways so that people can experience different circumstances and be curious about another space.

In upper space, white wall hides original wall and equipment like air conditioning system or piping. And the wall can serve as frames which show some graphic, plants and scenery of the restaurant. In addition, it surrounds stairs space and makes big void.

In lower space, you can slightly feel the border (plan) of white wall above your head and it's quite different from the plan of the restaurant floor partitioned by brick wall. The brick wall turns in right angle many times and makes hall space on the inside. And it also makes private room, kitchen and storeroom on the outside, which is between brick wall and original wall.

Design_ Chikara Ohno / sinato   **Photography_** Toshiyuki Yano   **Country_** Japan   **Client_** Dream Studio Co.,Ltd.

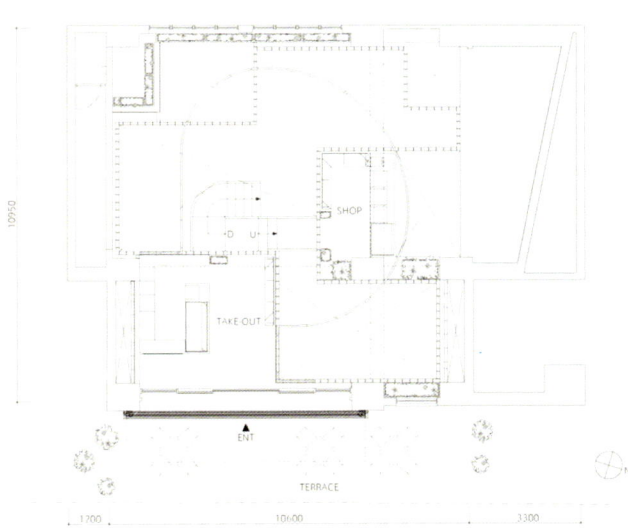

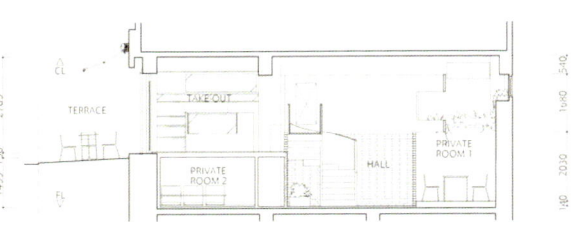

_UPPER LEVEL PLAN

_SECTION

WORK*shop* _ISSUE.TWO

WHAT HAPPENS

_The Metrics

What Happens When is a temporary restaurant installation that transforms every 30 days. Each month's theme we call a Movement. It will live for 9 months in a reclaimed space in New York City and will change focus every month in order to explore what a dining experience can be and how we can play with the traditional expectations of dining out.

The overall 'ork in progress' concept for the space is designed to reflect the changing and experimental nature of the project. With our actual architectural drawings projected onto the surfaces of the space in scale 1:1 the guests are invited into the design process. To serve as a backdrop for the monthly changes we inverted our drawings to give the functionality of a theatre black box. The ceiling is covered with a 12" grid of hooks to keep the space flexible and to be able to easily reconfigure the lighting for each Movement. All the ceiling lights have 15' cords.

Within this framework we design a new spatial concept for each Movement based on the theme. With only one night to do the transformations and our limited budget, our main tools for creating a new setting for each theme are lighting, colour scheme and spatial elements that can be prepared off site.

RESTAURANT/BAR | Design_ The Metrics | Photography_ Felix de Voss | Country_ USA | Client_ Private

Where the wild things are

With the forest / where the wild things are, theme of Movement 2, the interior design concept for Movement 2, is taking on a play with scale. The space is defined by an installation of over-sized pine needles that create a movement across the ceiling. Stretching to the floor in some areas the pine needles act as room dividers. Throughout the space little moments unveil the fantastical forest theme such as two moss-laden swings with miniature landscapes of plants and birds, bird houses nestled in the pine needles and various animal tracks on the floor, walls and selected tables. The over table pendants are made from a sheet of stationary held together with a single staple. The stationary, screen printed bird motif on vintage typing paper, is designed by Adrienne Wong. The counter area light fixtures are made with live moss and small bulbs.

Garden Party a la Renoir

Movement 3 takes on a spring garden party theme inspired by Renoir's 'The Luncheon of the Boating Party'. Drawing inspiration from this renowned Impressionistic painting, Movement 3 is about taking a trip into the 1880s France via 2011 New York.

The interiors draw a few significant elements and from the Renoir painting to re-create an intimate, communal experience reminiscent of 19th century time and place. A 25' awning like architectural stroke across the room frames the dining settings in warm spring like tones to re-create the communal intimate feeling of the painting.

The ceiling is lit with a 'garden party' string light inspired configuration of bulbs. Creating a play between interior and exterior, elements such as branch-like light fixtures and iconic still life wall sconces nod both to nature and the genre of painting. Throughout the space, guests will likewise discover details that reference the time period, such as 1900s inspired graphics across the tables and period pieces.

The bathrooms are divided between male and female interpretations of the time period, drawing inspiration from a boudoir and harbor graphics respectively.

Jazz

Inspired by Jazz, Movement 4 explores the tension between improvisation and tradition, as well as the unique rhythm and architecture that defines this musical form. A non-traditional take on jazz, this Movement Elle Kunnos de Voss uses 9,100ft of string to create different volumes throughout the space. This spatial rhythm throughout the room results in a visual cadence from the entrance all the way to the back that visually and spatially communicates the rhythm and contrast of Jazz. Amber tones of string hearken back to southern sunlight and remind guests of the birthplace of this unique American form of music.

The entryway welcomes guests with a taste of what's to come with a collage of patterns and imagery that mix old photographs from New Orleans with abstract patterns that visually represent the tempo of jazz. Adding another 'note' of reference, all the light fixtures are inspired by jazz line up instruments. Made from basic geometric shapes in bright colours, they are configured around the room at different heights, adding to the visual symphony.

Adding to the environment, the menu for Movement 4 is based off the birthplace of American jazz and is inspired by New Orleans flavors and techniques. Both playful and innovative in its use of traditional dishes like gumbo and po' boys, the food capitalises on the spice of these classic American recipes, equally delighting and exciting palettes while staying true to the simplicity of these flavors.

In compliment to the flavors and textures, guests are treated to the sounds of Diallo Riddle, a DJ who mixes New Orleans Jazz with modern notes, referencing the city and surrounding innovation of when this musical form was birthed.

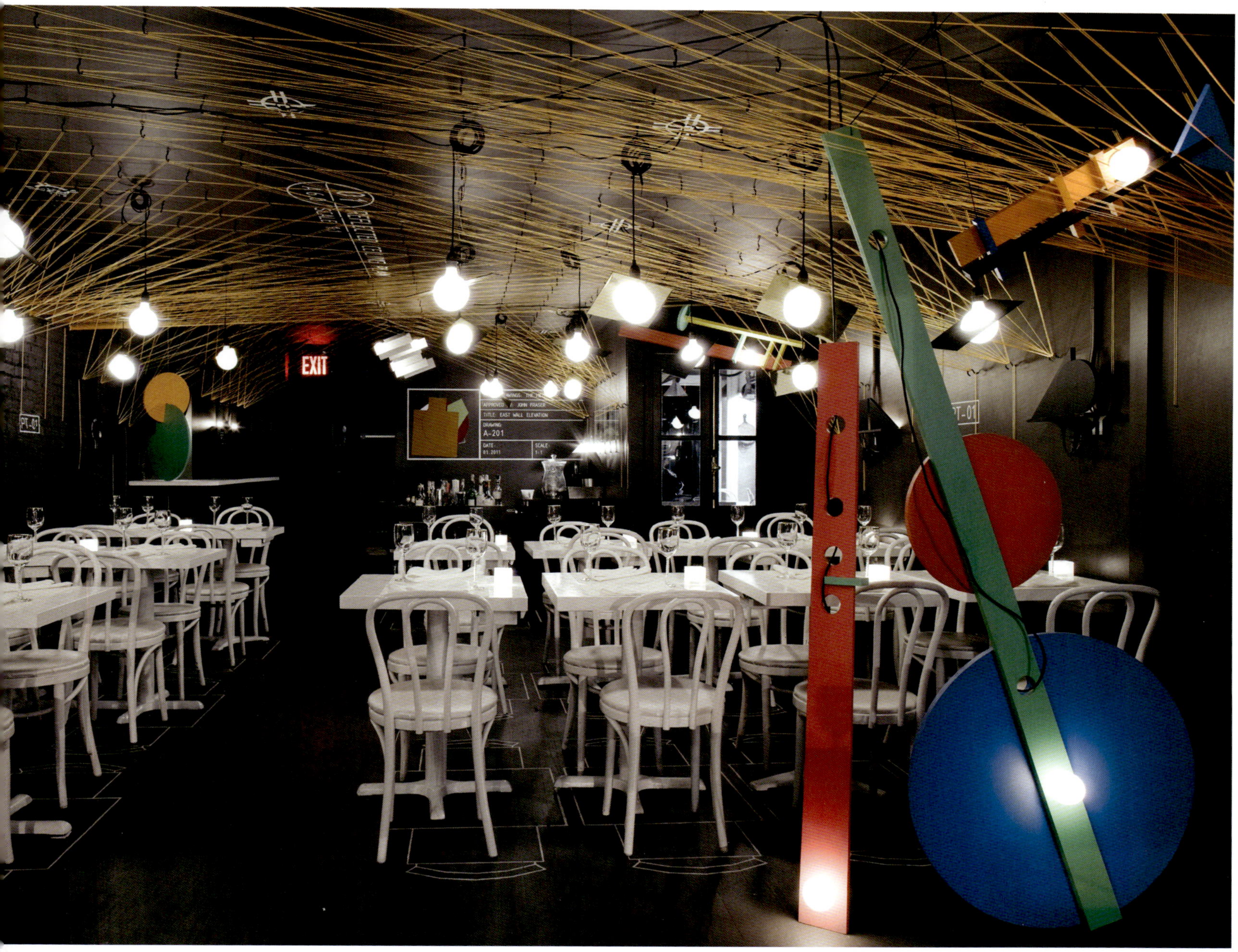
EXIT
PT-01
A-201

EXIT

don't flush

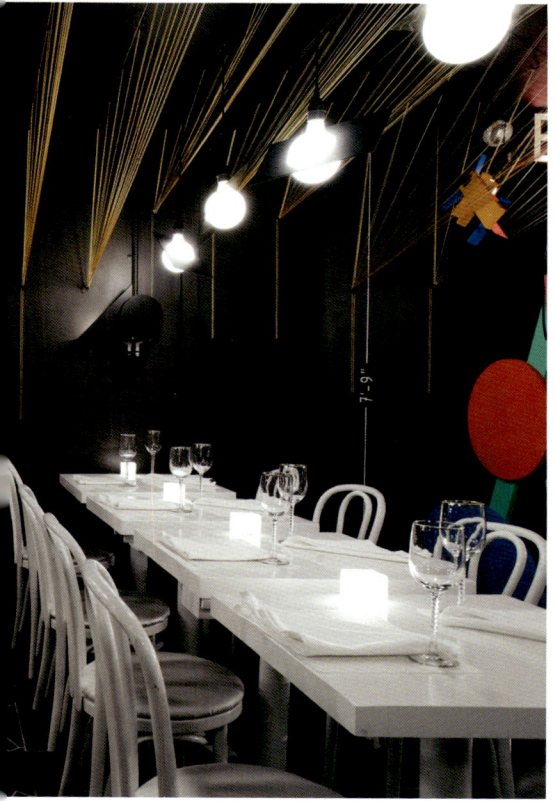

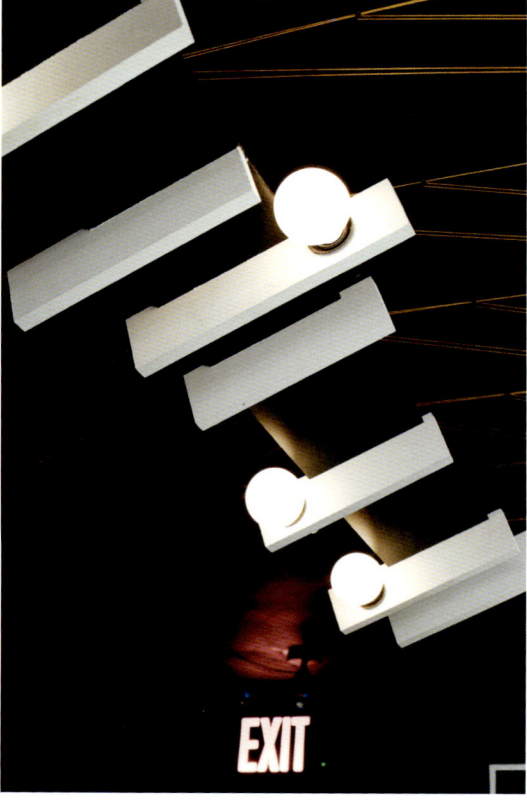

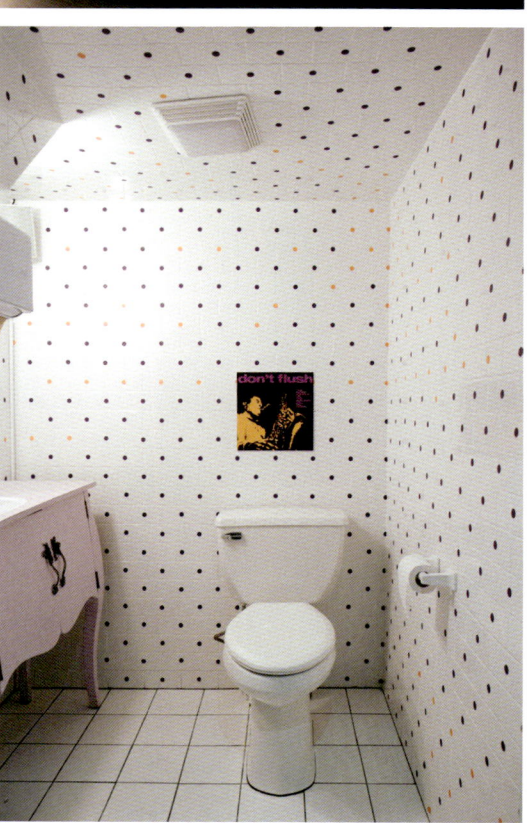

OFFICE

03

NO PICNIC

_Elding Oscarson

No Picnic is one of the world's largest design consultants, covering industrial design, product design and packaging design, as well as art direction, consumer insight, and architecture. We could hardly imagine a better oriented client, and expected nothing less than an ambitious, demanding, and fun project. They wanted large and open office spaces, a prototype workshop, a prototype showroom, several project rooms, and a striking customer area, distinctly separated from the other spaces in order to maintain secrecy.

For this, the client had found a group of 19th century buildings in central Stockholm, mainly consisting of two volumes, one originally an exercise hall for troops, and the other once a stable for police horses.

They had been converted into showrooms in the 1980s, and were in a sad state. These buildings currently enjoy the highest level of historical protection.

Conversion had to be sensitive, and we have evaluated every step with an antiquarian, literally down to each new screw hole.

We wanted to get rid of all added layers down to the origin. In the old stable we were able to peel the room naked, and just add a custom designed acoustical treatment along the walls, but in the exercise hall, economy and function demanded that a mezzanine constructed there in the 1980s, was kept. The mezzanine cut the hall lengthwise, and crippled the experience of the space in an unfortunate way.

OFFICE | **Design**_ Elding Oscarson | **Photography**_ Åke E:son Lindman | **Country**_ Sweden | **Client**_ No Picnic AB

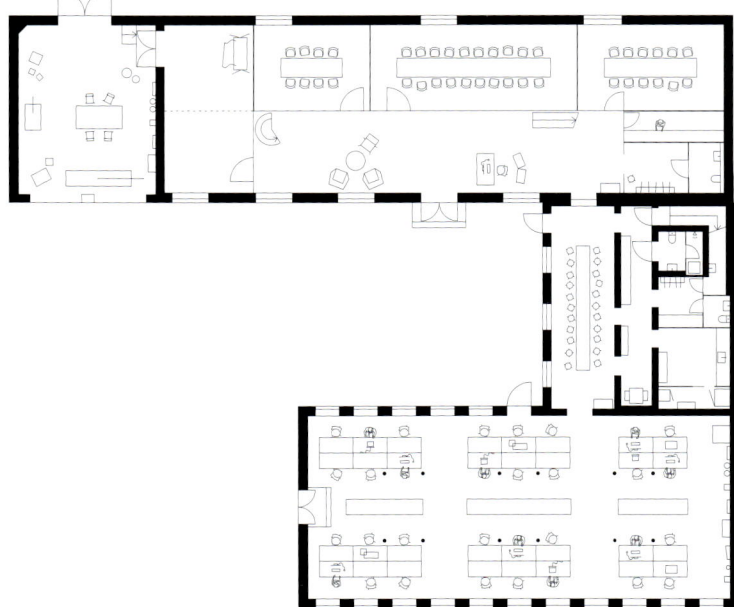

Its edge coincided with the centre of the hall, so we opted for the industrial designer's own method – the way arbitrary but symmetric shapes can be sculpted as half models onto a mirror, we could restore the impression of the entire exercise hall by constructing a delicate aluminum wall along its central axis. The meeting rooms inside this metal membrane, has large window panes towards the hall. The flat reflection of the glass appearing flush with the distorting metal surface, makes the glass seem like a mirror while the metal appears transparent; the wall is there, yet it disappears. It is bold, kaleidoscopic and delusive with its trompe-l'oeil effects. At the same time it takes a step back for the main act: the light and space of the exercise hall, and the old building's straightforward display of material, construction, imperfections, and time that has passed.

_GROUND FLOOR **PLAN**

OBSCURA DIGITAL HEADQUARTERS

_IwamotoScott Architecture

Obscura Digital is an expanding and well-recognised immersive and interactive media company with a cutting-edge artistic ethos. This project called for innovative architectural strategies to transform a typical office building into an artistic design and production headquarters. It also required creative construction and material techniques to meet the strict budget and schedule constraints.

Obscura's work involves design, research and development, fabrication and installation across a range of digital media, including architectural projections and large interactive touchscreens. Programmatic requirements for their space include office and meeting spaces for the partners and staff, a large multi-functioning showroom and exhibition area, workstations for digital production, and prototyping workshop spaces.

The new space is within a three-storey 1940s concrete and steel frame warehouse in the Dogpatch neighborhood of San Francisco, shared with IwamotoScott's office. The main programmatic elements are distributed with the prototyping workshop and showroom on the lower level, reception, main offices and meeting spaces on the middle level, and production on the top level.

A main aspect of the redesign of this warehouse involves the removal of three bays at the centre of the building to form a spatial connection wbetween the lower and mid levels. This move also allows the installation of one of Obscura's hemispherical projection theatres housed within a 30' diameter geodesic dome.

Overlooking this new double height space is a central conference room designed as a freestanding object peering over the end of the new void, oriented towards the dome. One side of this 'meeting box' is kinked inward to adapt to existing seismic bracing and allow a communication stair to slip alongside that connects the two levels at the edge of the void. This meeting box is treated as sculptural object with different interior and exterior materialities.

A broad new stair at the entry allows an easy spatial flow into the reception area, partners' offices and central meeting space. Partner and staff offices are defined by twisting polycarbonate screen walls that allow filtered views, as well as the transmission of natural light from perimeter windows and skylights above.

_ENTRY

OFFICE | Design_ IwamotoScott Architecture | Photography_ Rien van Rijthoven | Country_ USA | Client_ Obscura Digital

_CONFERENCE **ROOM**

_IWAMOTOSCOTT ARCHITECTURE **OFFICE**

_PROTOTYPING **AREA**

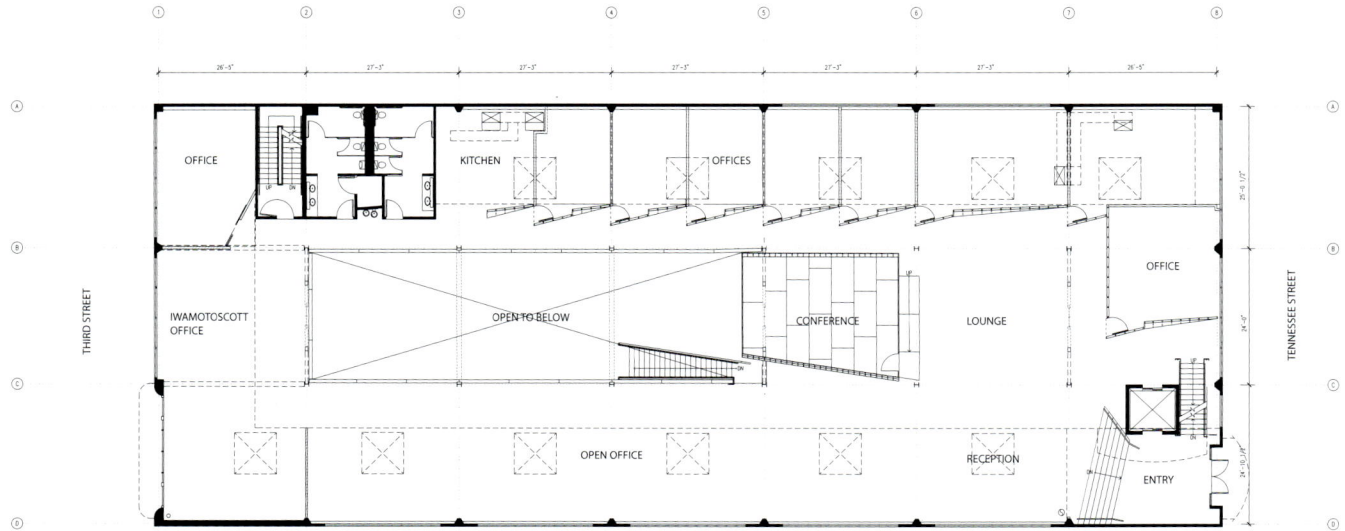
_MAIN FLOOR **PLAN**

BAJAJ CORPORATE OFFICE IN INDIA

_Collaborative Architecture

The design for Bajaj Electricals corporate office is an attempt to take the project above the functional contingencies of a regular corporate work place brief, imbibing the space with an architectural character that would generate unparallel work efficiency.

The project had typical straight jacketed, hierarchical brief for a corporate office, with cabins for senior managers, cubicles for mid-level managers and workstations for executives, meeting room conference room, staff area and service areas. The brief was also emphasised to have orthogonal spaces in the office, in line with traditional Indian 'Vastu' principles.

Sustainability as an active strategy was fundamental theme right from the inception of concept design through selection of green rated products, active tools to cut down the energy consumption, day light harvesting and the general orientation of the work floor to tap maximum day light and reduce the south exposure to cut down cooling load.

The design challenge was to accommodate these highly debilitating guidelines, yet evolve a 'design narrative' which will cleverly mask the ubiquitous gridded plan. Meeting room is placed strategically to segregate the public interface with the work interface. The contrasting dark colour and the free flowing envelope of its space clearly generate a spatial boundary between Public-Work realms.

Enclosed spaces like senior manager cabins, conference, staff utility areas are positioned along one of the sides, allowing the workstations to be positioned to tap maximum day lighting.

The cabins have glazed partitions separating them from the workspace, allowing the day light to penetrate in to the cabin spaces. Collaborative designed the architectural graphics on the wall to ensure adequate privacy in the cabin area.

Every cabin, passages and entry foyers are provided with active tools to cut down the lighting load. The office is equipped with state of the art HVAC and BMS to make it truly energy efficient.

OFFICE | **Design_** Collaborative Architecture | **Photography_** Lalita Tharani | **Country_** India | **Client_** Bajaj Electricals Ltd.

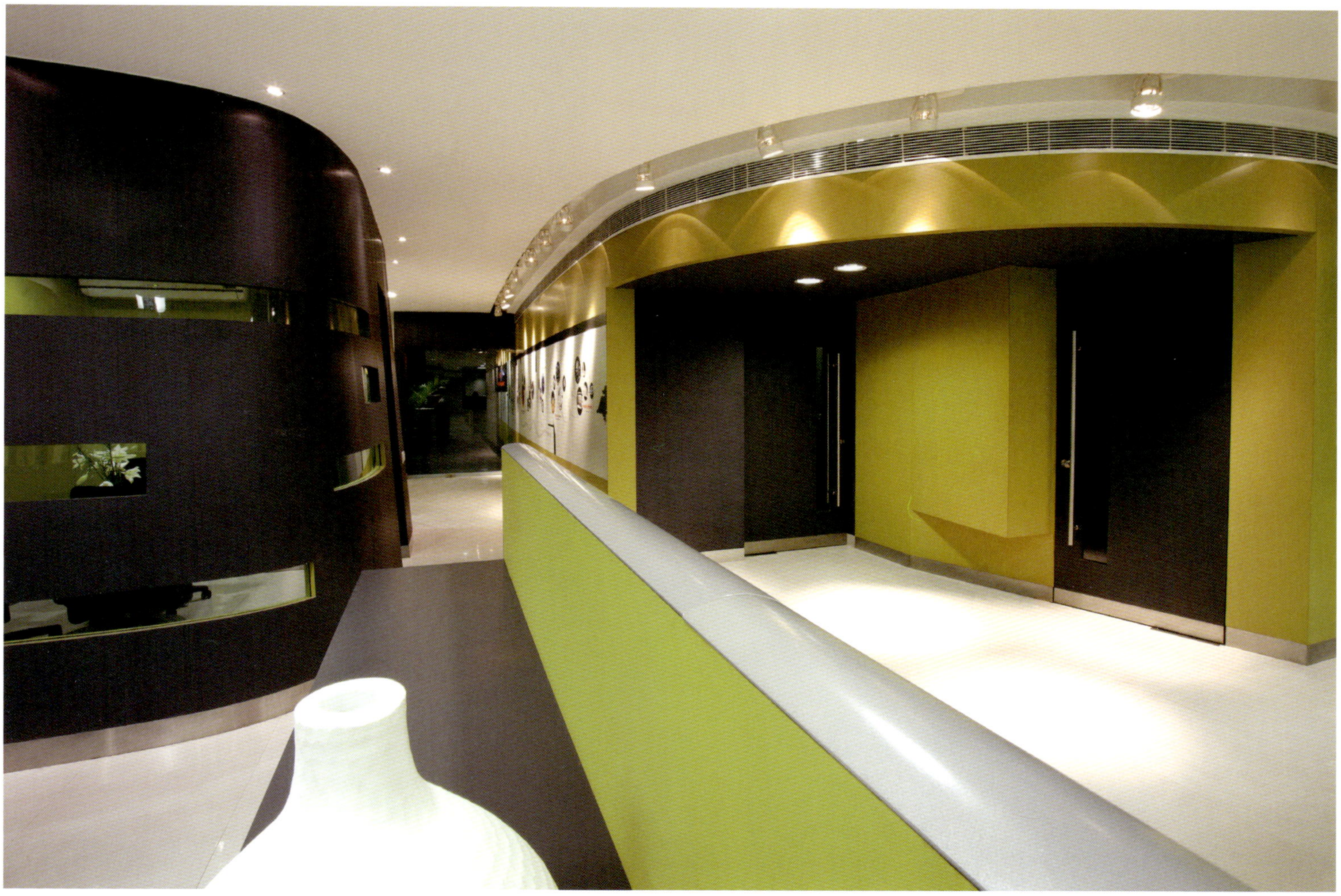

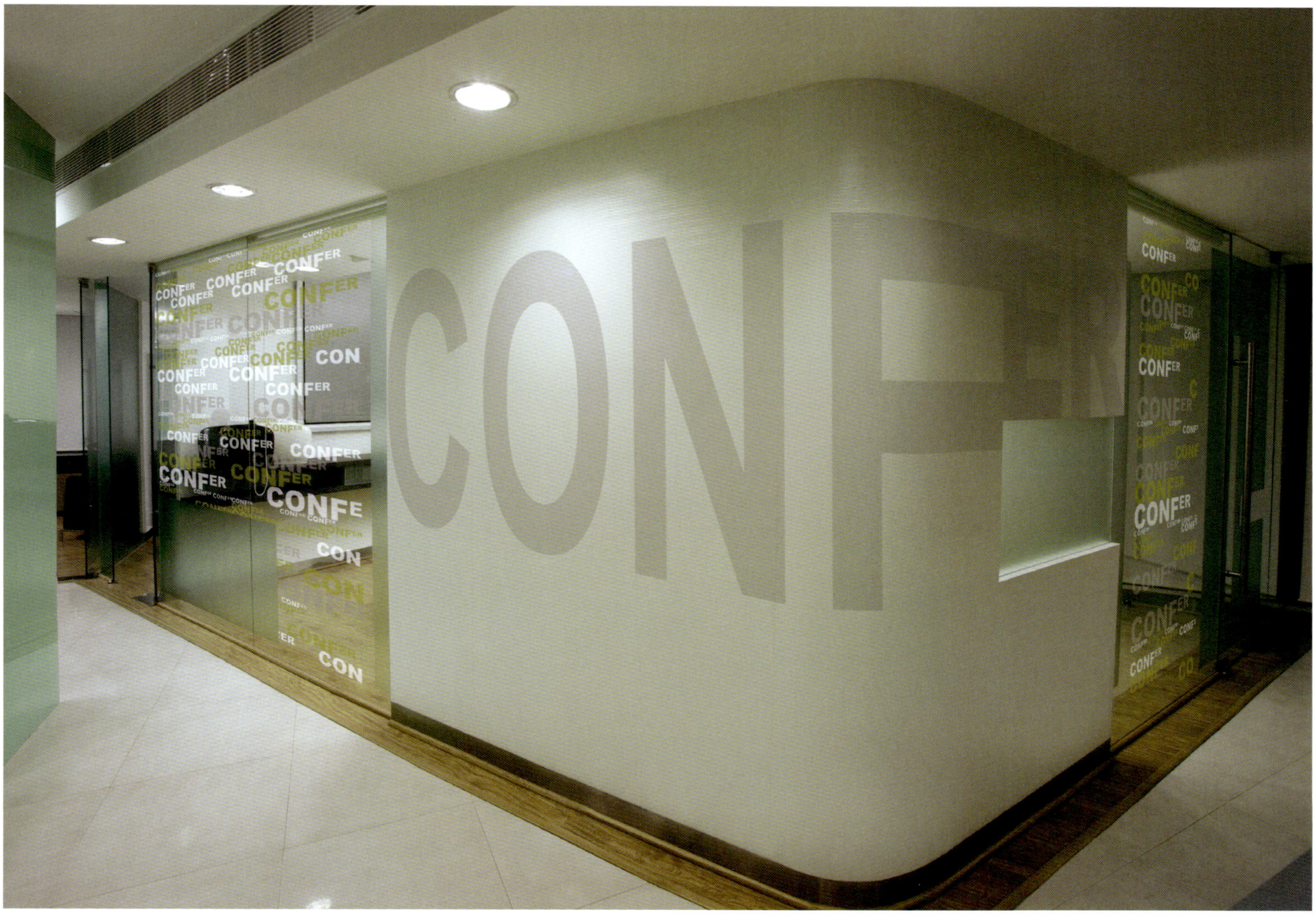
CONFER

WORK*shop* _ISSUE.TWO

SUGAMO SHINKIN BANK: TOKIWADAI BRANCH

_emmanuelle moureaux architecture + design

Sugamo Shinkin Bank is a credit union that strives to provide first-rate hospitality to its customers in accordance with its motto: 'we take pleasure in serving happy customers.' We handled the architectural and interior design for the bank's newly relocated branch in Tokiwadai.

By basing our design around leaf motifs, we sought to create a refreshing space that would welcome customers with a natural, rejuvenative feeling. The façade of the building features silhouettes of trees and an assortment of both large and small windows in 14 different colours arranged in a distinctive, rhythmical pattern that transforms the façade itself into signage.

| OFFICE | Design_ emmanuelle moureaux architecture + design | Photography_ Nacasa & Partners Inc. | Country_ Japan | Client_ Sugamo Shinkin Bank |

WORK*shop* _ISSUE.**TWO**

ATMs and teller windows are located on the first floor, along with 3 courtyards and an open space laid out with chairs in 14 different colours. The second storey houses the loan section, reception rooms, offices and 4 courtyards, while the third floor is reserved for facilities for staff use, including changing rooms and a cafeteria. Thanks to the 7 light-filled courtyards planted with trees and flowering plants, each of these spaces is loosely connected to all the others. A constellation of leaves in 24 different colours growing on the white branches of the walls and glass windows overlaps with the natural foliage of the real trees in the courtyards, creating the sensation of being in a magical forest.

SUGAMO SHINKIN BANK: SHIMURA BRANCH

_emmanuelle moureaux architecture + design

OFFICE | **Design**_ emmanuelle moureaux architecture + design | **Photography**_ Nacasa & Partners Inc. | **Country**_ Japan | **Client**_ Sugamo Shinkin Bank

_SUGAMO SHINKIN BANK: **SHIMURA BRANCH** _OFFICE _111

Sugamo Shinkin Bank is a credit union that strives to provide first-rate hospitality to its customers in accordance with its motto: 'we take pleasure in serving happy customers.' Having completed the design for branch outlets of Sugamo Shinkin Bank located in Tokiwadai and Niiza, we were also commissioned to handle the architectural and interior design for its newly rebuilt branch in Shimura. For this project, we sought to create a refreshing atmosphere with a palpable sense of nature based on an open sky motif.
ATMs, teller windows, consultation booths and an open space laid out with chairs in 14 different colours are located on the first floor.
The second storey houses offices, meeting rooms and a cafeteria, while the third floor is reserved for the staff changing rooms.
Three long glass airwells thread through the first and second levels of the building, flooding the interior with natural light as well as 'blowing' air through it.

12 Layers of Colours
A rainbow-like stack of coloured layers, peeking out from the façade to welcome visitors.
Reflected onto the white surface, these colours leave a faint trace over it, creating a warm, gentle feeling. At night, the coloured layers are faintly illuminated. The illumination varies according to the season and time of day, conjuring up myriad landscapes.

A Piece of the Sky
Upon entering the building, three elliptical skylights bathe the interior in a soft light. Visitors spontaneously look up to see a cut-out piece of the sky that invites them to gaze languidly at it. The open sky and sensation of openness prompts you to take deep breaths, refreshing your body from within.

Fuzzy Puffs
The ceiling is adorned with dandelion puff motifs that seem to float and drift through the air. In Europe, there is a long and cherished custom of blowing on one of these fuzzy balls while secretly making a wish. Bits of fluffy down gently dance and frolic in the air, carried by the wind.

DE NIJE GRITENIJE

_FLATarchitects

The Nije Gritenije foundation is situated on the top floors of the new Rabobank Heerenveen-Gorredijk head office. FLATarchitects designed the interior containing twenty workstations in a flexible working environment.

The Nije Gritenije foundation is a Rabobank initiative to stimulate local and regional entrepreneurship. The foundation asked Amsterdam based FLATarchitects to design twenty workstations and a conference room on the ninth and tenth floor of the brand new Rabobank building. Four of these workstations will be permanently used by members of the foundation. The remaining sixteen workstations will be used by a continuously changing group of starting entrepreneurs and artists.

FLATarchitects devised a basic interior design which can be customised and completed by the users. They can adapt their space to their specific needs. In this way temporary users are able to appropriate their place of work in a relatively short period of time. Also, the interior is designed to encourage innovation and discussion among users and visitors.

The stairs towards the ninth floor end in an entrance space fenced off by four irregularly shaped objects. Together they form a 'porous wall' filtering the activity in the work space behind. Through the gaps and reflecting surfaces of the wall, sounds and images are presented to the spectator in fragments. The shifting sight which lines of the visitor climbing the stairs was used as a designing instrument for these wooden objects. These same objects on the side of the work space are used as bookshelves. The permanent workstations are positioned next to the entrance. The four desktops fan out from a shared drawer unit. Their shapes and colours differ from the flexible workstations. Four large tables together provide space for sixteen temporary workers. The tables consist of a steel frame with a tabletop made of steel grid panels. The tabletop can be arranged according to one's needs by combining various desk elements, such as desktops of several sizes, storage objects, book displays and multiple sockets. These elements have a negative grid at the bottom, in order to fix them to the grid.

OFFICE | Design_ FLATarchitects | Photography_ Arend Loerts | Country_ The Netherlands | Client_ Rabobank

From the entrance space stairs lead on to the conference and presentation room on the tenth floor. In order to emphasise the relatedness of the two top floors, the striking banisters by their irregular shape refer to the wall objects. Upstairs your attention is immediately drawn to the carpet, which is also covering parts of the wall. It shows a giant map of Friesland (the Dutch province) designed by graphic artist Martin Draax. The members of the Nije Gritenije foundation gather over this map, around a grand conference table with eleven chairs, one for each Frisian city. By closing a wall-to-wall curtain, part of the space can be dimmed for showing films or presentations. When looking out of the windows the real map of Friesland unfolds.

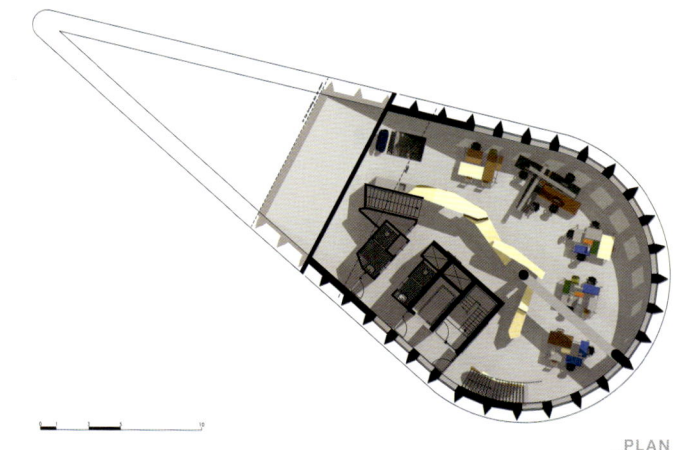

_PLAN

WORK*shop* _ISSUE.TWO

INFO DESK IN REAR **LOUNGE**

OFFICE Design_ NAU, DGJ Photography_ Jan Bitter Country_ Germany Client_ Raiffeisen Schweiz

_LOBBY

_CNC MILLED **PORTRAITS**

_CONFERENCE **ROOM**

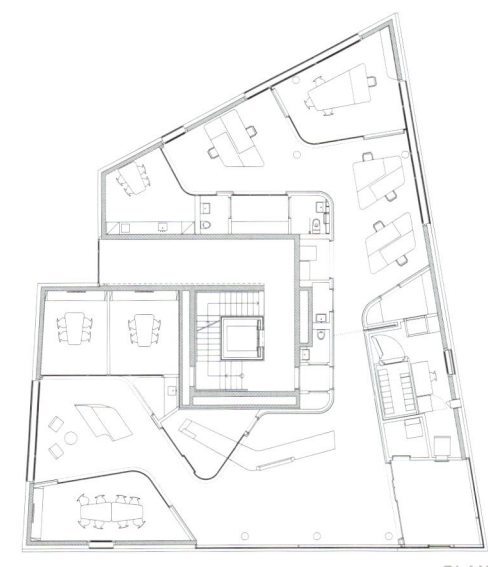
_PLAN

OPEN LOUNGE

_NAU, DGJ

Raiffeisen's flagship branch on Zurich's Kreuzplatz dissolves traditional barriers between customer and employee, creating a new type of 'open bank', a space of encounter. Advanced technologies make banking infrastructure largely invisible; employees access terminals concealed in furniture elements, while a robotic retrieval system grants 24 hour access to safety deposit boxes. This shifts the bank's role into becoming a light-filled, inviting environment – an open lounge where customers can learn about new products and services. This lounge feels more like a high-end retail environment than a traditional bank interior. Conversations can start spontaneously around a touchscreen equipped info-table and transition to meeting rooms for more private discussions. The info-table not only displays figures from world markets in realtime, but can be used to interactively discover the history of Hottingen, or just check the latest sports scores.

Elegantly flowing walls blend the different areas of the bank into one smooth continuum, spanning from the customer reception at the front, to employee workstations oriented to the courtyard. The plan carefully controls views to create different grades of privacy and to maximise daylight throughout. The walls themselves act as a membrane mediating between the open public spaces and intimately scaled conference rooms. Portraits of the quarter's most prominent past residents like Böklin, Semper or Sypri grace the walls, their abstracted images milled into Hi-macs using advanced digital production techniques. While intricately decorative, the design ground the bank in the area's cultural past, while looking clearly towards the future.

WORK*shop* _ISSUE.TWO

THIN OFFICE

_Studio SKLIM

| OFFICE | Design_ Studio SKLIM | Photography_ Jeremy San | Country_ Singapore | Client_ Kido Technologies |

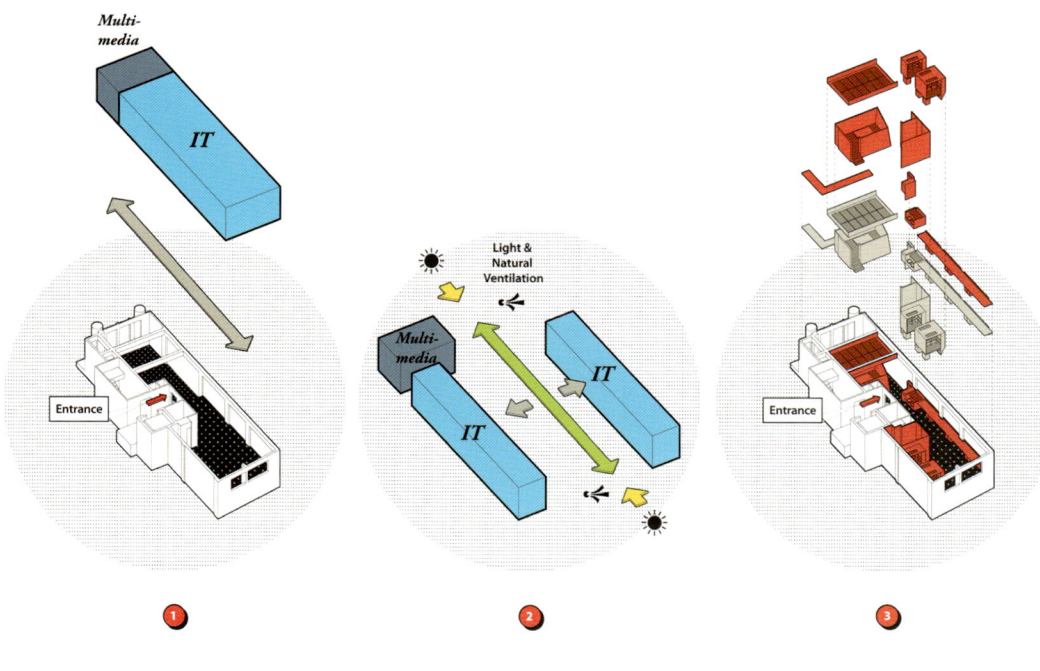

_CONCEPT

_MAIN **THOROUGHFARE**

Thin office provides a blank canvas for two offices to operate amidst flexible working facilities and casual niches.

While tapping on a laptop in a café has become the ubiquitous platform to begin 'work', the need for a permanent work environment for any office is still necessary in the long run. Perhaps what has changed since the advent of 'coffee offices' has been the increasing need for flexibility within a sedentary work sphere.

The program brief was for an office space shared by an IT company and a multi-media setup. Located in a refurbished postwar building right in Singapore's CBD outskirts, the space was long and narrow with split levels, offering the possibility of a raised space. Throughout the long and narrow office, the ceiling and wall conditions were left unaltered as much as possible, along with the existing light fixtures.

The designed space was to reflect the ethos of the companies: Flexibility, Technology and Creativity. The office space was loosely organised into 8 clusters namely: the Boss Boxes, Long Work Top, Discussion Table, Welcome Mat, Sanitary & Storage, Recharging Point, Twist Platform and Multi-media Corner. Each of these clusters was arranged around an open plan configuration with the exception of Sanitary & Storage to allow a multifarious overlap of working trajectories.

The flexible working environment was kept in mind with the possibility of hot-desking, informal working clusters and also semi-private cubicles. The Boss Boxes were an option for more privacy as some work required a certain level of seclusion. Technology is a crucial aspect of any modern day office and the ease of being 'connected' to

WORK*shop* _ISSUE.TWO

_TWISTING **PLATFORM**

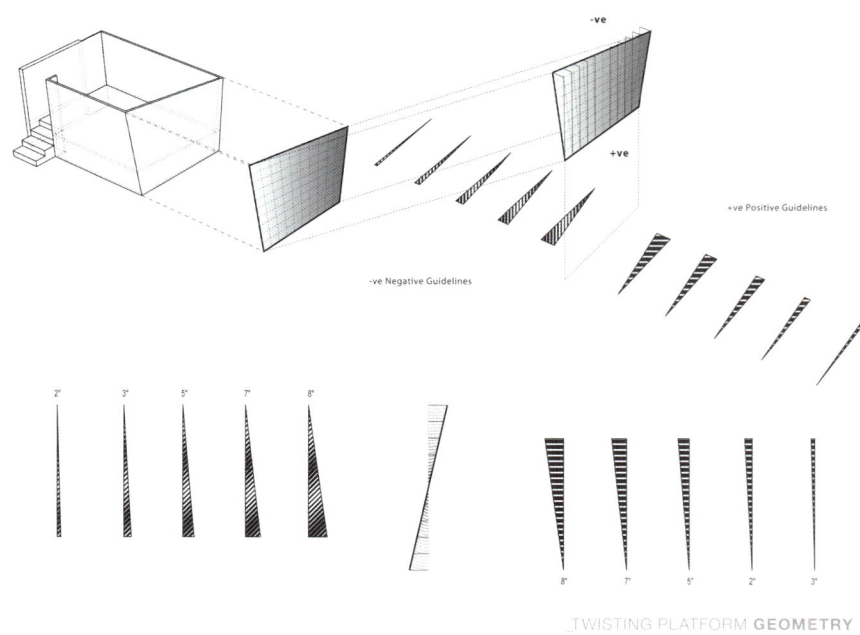

TWISTING PLATFORM **GEOMETRY**

either an internet network or a power source was one of the concerns of the client. The fluctuating size of the workforce also meant flexible working spaces which could be contracted and expanded to fit the demands of this office. The result was the 'Long Work Top' which incorporated an ingenious power strip of data points, power supply and telecommunication points to be accessible at any location along this table, expanding the number of workstations from 6 to 10 in a few minutes! This single piece of stretched work surface became part of a greater string of furniture transforming from table top, reception seating, storage and finally to pantry space.

The Twist Platform was a raised meeting pod that capitalised on the higher ceiling to incorporate storage beneath. The geometry of the subtly twisting space was driven by sightlines, privacy and anthropometrics. The unconventional form in an otherwise sleek and straightforward office space added a dynamic backdrop to the Recharging Point and provided privacy to the independent operation of the multi-media setup. The giant overhead light fixture was a final touch to the suggestion of this event space.

The essence of this 'Thin Office' was a desire to remain anonymous and to provide a blank canvas for various work scenarios and possibilities. This 'thinness' was translated from the basic organisation of spaces which opened up a central thoroughfare for circulation, light and natural ventilation, through to the furniture details which celebrated the geometrical state of being folded, suspended or twisted.

WORK*shop* _ISSUE.TWO

STUDIO SC

_Studio MK27

The architectural project of this photography studio, specialised in food photos, emerged from an internal competition held at Studio MK27. The team was divided into 3 groups that worked on the development of different ideas for one day. From these first rough sketches a new project, which, in part, was a synthesis of all these sketches and, in part, an entirely new project, was elaborated.
In the definitive design, the land was longitudinally divided into two. The Northern part, with a width of 7.2m and a length of 43.5m, was reserved for a generous garden. This space functions as an internal esplanade for the building. The Southern part, with a width of 12.2m and a length of 43.5m, in turn, was totally built and contains the program of the studio.

OFFICE | **Design_** Studio MK27 | **Photography_** Nelson Kon | **Country_** Brazil | **Client_** Studio SC

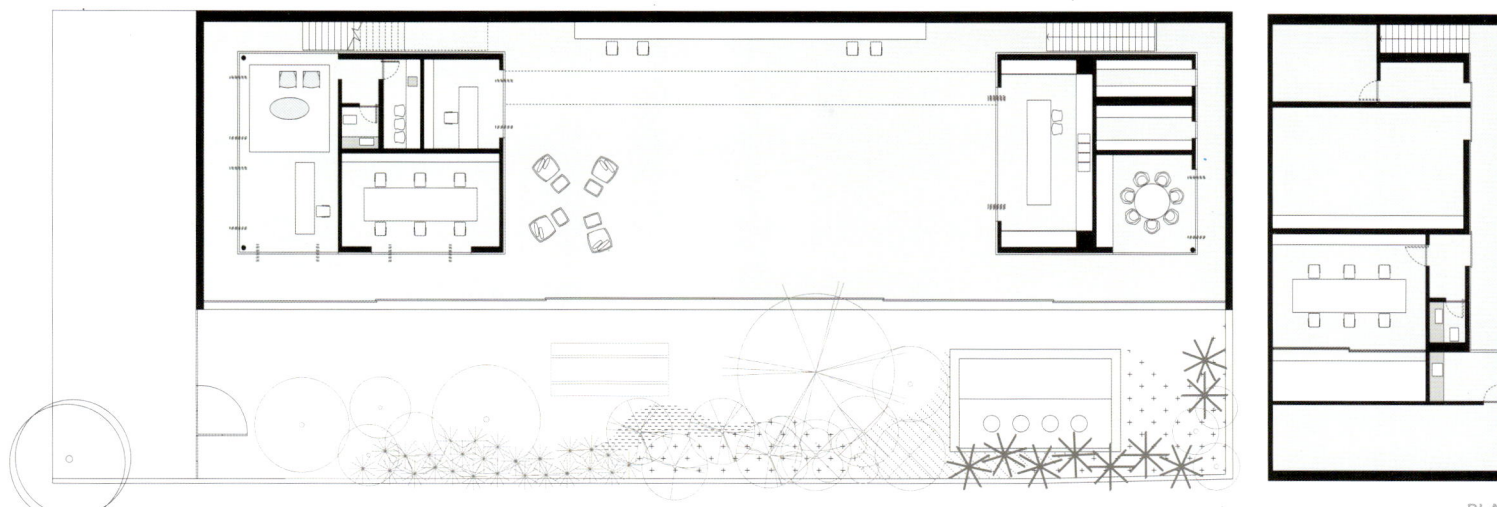

ground floor 1_200

_PLAN

_CONCRETE WALKWAY

_WOODEN BOX

_RECEPTION

_KITCHEN

The entrance is via the garden: large sliding metal doors open in their entirety and create a total continuity between the central emptiness and the external space. The main space of the studio is cut by a concrete walkway suspended from the ceiling. This element connects two wooden boxes and configures an internal overlook. On the ground floor, the first wooden volume, closer to the parking lot, is the reception area and a room for image treatment and, on the first floor a work-room lit by an internal patio. The second volume, on the ground floor, houses storerooms and a technical kitchen, which prepares the food for the photos and, on the first floor, an image treatment room.
On the top floor, in addition to a wooden deck, there is a social area for receptions. It is a large open kitchen from which chefs can prepare complete meals which are served right on the counter-top. One of the initial principles of the architecture for the building was to create a complete trajectory, going through the walkway and the main spaces of the studio before the visitor arrives at the upper kitchen. In this way, he is familiar with the work places, even if the studio is not working.
For this project, Studio MK27 sought to use industrial materials and installations. All the external finishings are of metal and, in the internal space there is a blend between metal and wood, which warms and affects the environment.

_STUDIO SC _OFFICE _125

_WALKWAY **SUPERIOR**

SIDEWALL

TELEVISOR

_WWAA + 137Kilo

The new Televisor's seat is located in the 90s building originally functioning as a single-family house. In order to adapt the building to its new, office role, the changes into its functional arrangement, technical infrastructure and the way of finishing had to be applied. At the same time the landlord of the premises required to be guaranteed a possibility of restoring the present state of the interior, especially a 'palace-like' parquet floor, after the termination of the lease agreement.

This issue together with the differences in the floor level, reaching over 70cm, of the whole groundfloor area, became a starting point for working out the concept of this design. Consequently, it is the flooring that acts as a basic interference element and gives the special character to the whole place. It has been constructed out of intersecting wooden planks laid at different angles. Being reminiscent of a scate park it creates a strong, dynamic effect, but simultaneously introduces some order into the space.

The new floor is laid on its own under-construction without removing the original parquet. In the place where it borders with the walls a gap of about 10cm was left to be filled with lighting conducted along its length. It produces an effect of a 'piece of furniture' inserted into an existing structure, light and softly illuminated at its edges.

Each storey of the building is of a different character and plays a different role. The most imposing groundfloor is open to the customers, the remaining storeys are more private.

The characteristic pattern of the floor as well as its natural oak-wood colours represent the predominant feature of the groundfloor area. The most significant elements of the groundfloor space, those defining the functions of their surrounding, are also directly connected with the floor, they 'grow out of it' in a material and a geometric sense, e.g., the staircase leading onto the next storey, a kitchen worktop, a reception desk and furniture in the lower part of a lounge. The central point of the lounge – a fireplace which opens on three sides – has also been encased in wood which turns up onto the ceiling and becomes a dominant feature but also conceals the main light source of the lounge.

OFFICE | Design_ WWAA + 137Kilo | Photography_ WWAA | Country_ Poland | Client_ Televisor

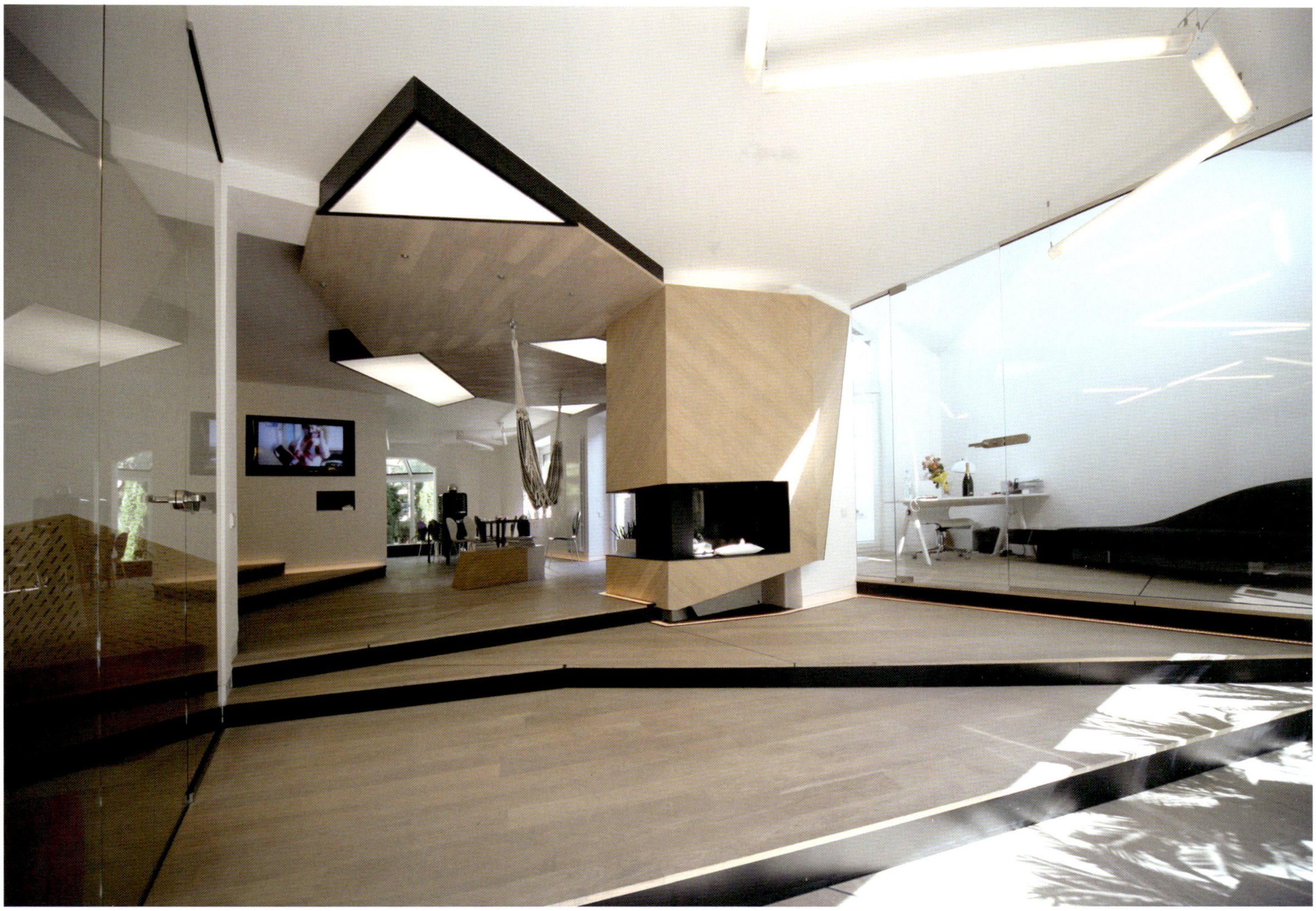

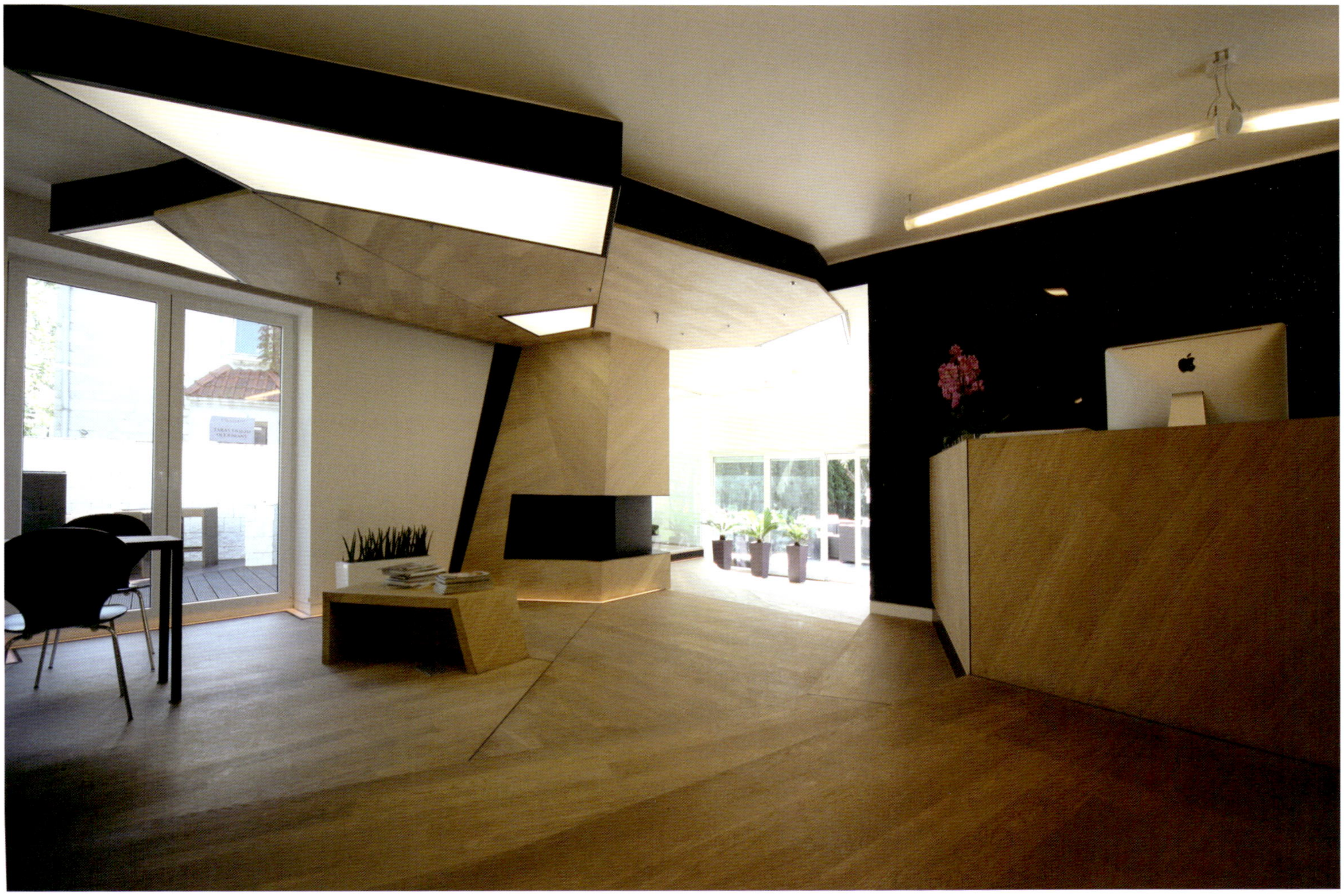

_PLAN

TRIBAL DDB AMSTERDAM

_i29 | interior architects

Tribal DDB Amsterdam is a highly ranked digital marketing agency and part of DDB international, worldwide one of the largest advertising offices. i29 interior architects designed their new offices for about 80 people. With Tribal DDB our goal was to create an environment where creative interaction is supported and to achieve as much workplaces as possible in a new structure with flexible desks and a large open space. All of this while maintaining a work environment that stimulates long office hours and concentrated work. As Tribal DDB is part of an international network a clear identity was required, which also fits the parent company DDB. The design had to reflect an identity that is friendly and playful but also professional and serious. The contradictions within these questions, asked for choices that allow great flexibility in the design.

Situated in a building where some structural parts could not be changed it was a challenge to integrate these elements in the design and become an addition to the whole. i29 searched for solutions to various problems

OFFICE   Design_ i29 | interior architects   Photography_ i29 | interior architects   Country_ The Netherlands   Client_ Tribal DDB Amsterdam

which could be addressed by one grand gesture. At first a material which could be an alternative to the ceiling system, but also to cover and integrate structural parts like a big round staircase.
Besides that, acoustics became a very important item, as the open spaces for stimulating creative interaction and optimal usage of space was required.
This led us to the use of fabrics. It is playful, and can make a powerful image on a conceptual level, it is perfect for absorbing sound and therefore it creates privacy in open spaces. And we could use it to cover scars of demolition in an effective way. There is probably no other   material which can be used on floors, ceiling, walls and to create pieces of furniture and lampshades then felt. It's also durable, acoustic, fireproof and environment friendly. Which doesn't mean it was easy to make all of these items in one material!

RABOBANK NEDERLAND

_Sander Architecten bv

Amsterdam firm Sander Architecten completed the Square of the office for Rabobank Nederland in Utrecht. Rabobank selected Sander Architecten out of a group of twenty to create and supervise the execution of the entire interior design (56,000m²), including the twentyfive-storey building. As the office interior is being redefined by the introduction of new methods of working, interior architecture is facing new challenges. In today's work environment, the emphasis is on cooperation in teams and group dynamics, people go to the office for the social aspect more than anything else. To realise this ambition, we view the building as a modern city. After all, the city is where individual freedom and spontaneous interaction are all-important.

Activity Based Architecture
The effectiveness of this concept is visible on the Square, located at the plinth of the new office building. Employees and visitors work, eat, read, and meet one another in a diverse landscape. The 'buildings', separate spaces with different functions, join up with the uncluttered grid of skylights and slim columns. The new style of working is based on freedom, trust and taking responsibility. In the client's view, its employees are all entrepreneurs, responsible for their own performance in an environment free of fixed rules, fixed times and fixed locations. The work spaces are tailored to specific activities: multi-person meetings, face-to-face meetings or a place to write a report with maximum concentration. Each activity has its own space.

To enable flows, vertical partitions were avoided so that the horizon would always be visible.

Form Follows Flow

In nature routes are formed naturally; people intuitively find their way. Architect Ellen Sander was seeking that naturalness, that 'flow'. The busiest routes automatically formed around the cores with the lifts and staircases, beyond which more peaceful zones naturally emerged. Moreover, the psychological concept of 'flow', the moment when need, desire and ability come together, connects the employee's sense of happiness with an optimum result for the employer. The guiding principle for the interior design therefore became 'form follows flow'. To enable flows vertical partitions were avoided so that the horizon would always be visible. 'The office is my world and the world is my office.'

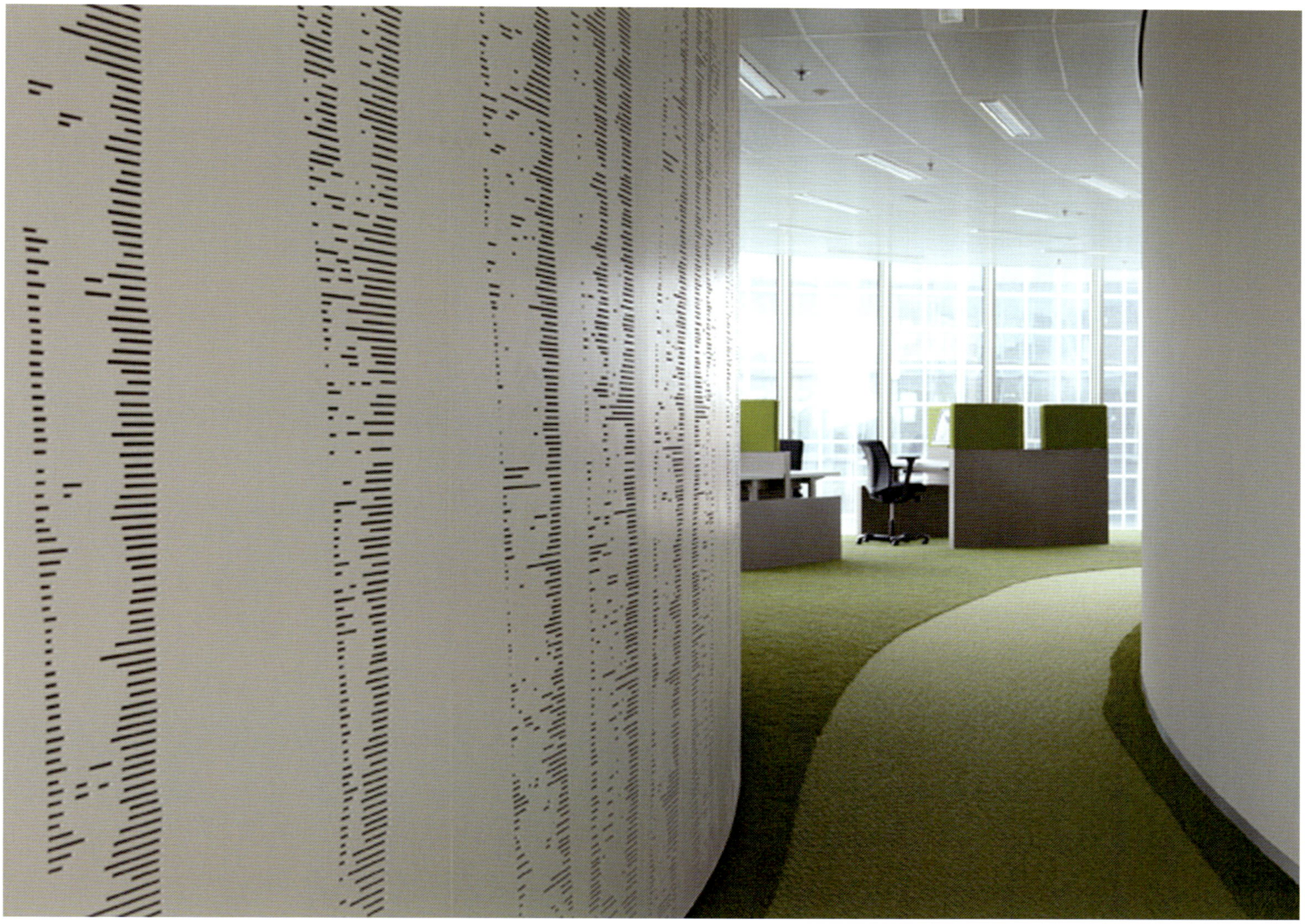

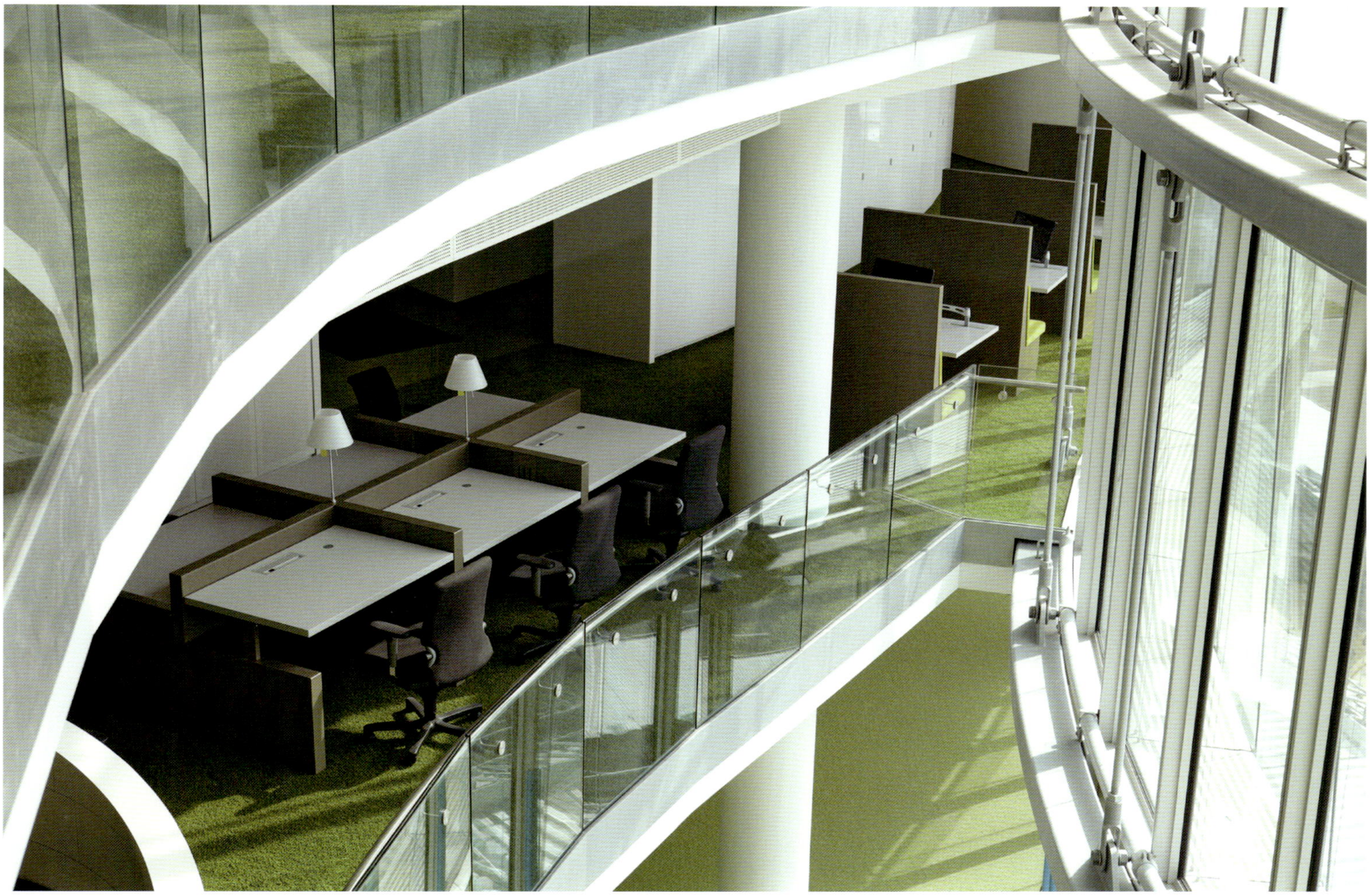

The paperboard pavilion, which features attractive patterns created by the different uses of the material, is particularly inviting to touch.

The design was generated by cooperation with a number of other designers. The Square on the plinth could not turn into a monotone, homogenous space. Diversity is required in order to stimulate people, and despite the enormous scale of the building, people are not left wandering around lost in sterile areas.
The meeting pavilions designed by Sander Architecten are made from washi paper and paperboard. In combination with the Chinese lantern from washi paper suspended from the skylight, a distinctively tactile experience is created. The paperboard pavilion, which features attractive patterns created by the different uses of the material, is particularly inviting to touch.

WORK*shop* _ISSUE.TWO

DDB IN SINGAPORE

_BBFL

As one of the world's leading advertising agency, DDB have found a new home at Pico Creative Centre in Singapore occupying 2 floors of 1,000M² each. BBFL were appointed as the architects to understand, interpret and create a workspace that embodies DDB's dynamic working culture.

The design process began with observing the designers at DDB work as a diverse group which involves a high level of discussion, debate and intervention sessions amongst each team with a common goal of achieving a creative solution for their clients. Group discussion activity was crucial for the team where workspace becomes a social network area rather than a confined individual space. The design process has allowed us to create a bridge between two disciplines which played a huge role in the concept development of the entire DDB environment creating a conspicuous identity.

The workspace concept that revolves around DDB's office wasn't just about its corporate colours or even the branding of its company but create a collective space that leans towards the culture of the people in DDB itself. This leads us to create spaces which improves and promotes inter-connectivity between creative individuals. To enhance this notion, spaces within the work environment weren't crafted out with walls or any form of concealed demarcation, compositing to an open office setting.

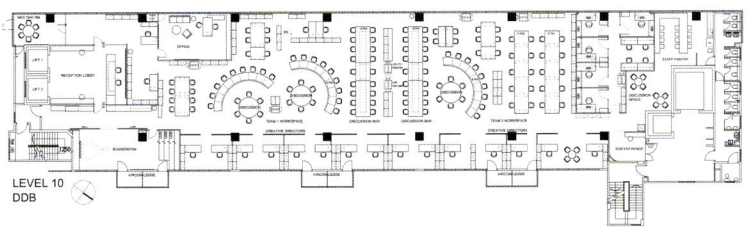

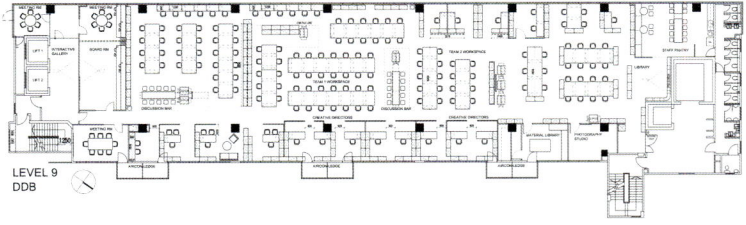

_PLAN

OFFICE | **Design_** BBFL | **Photography_** Living Pod | **Country_** Singapore | **Client_** DDB

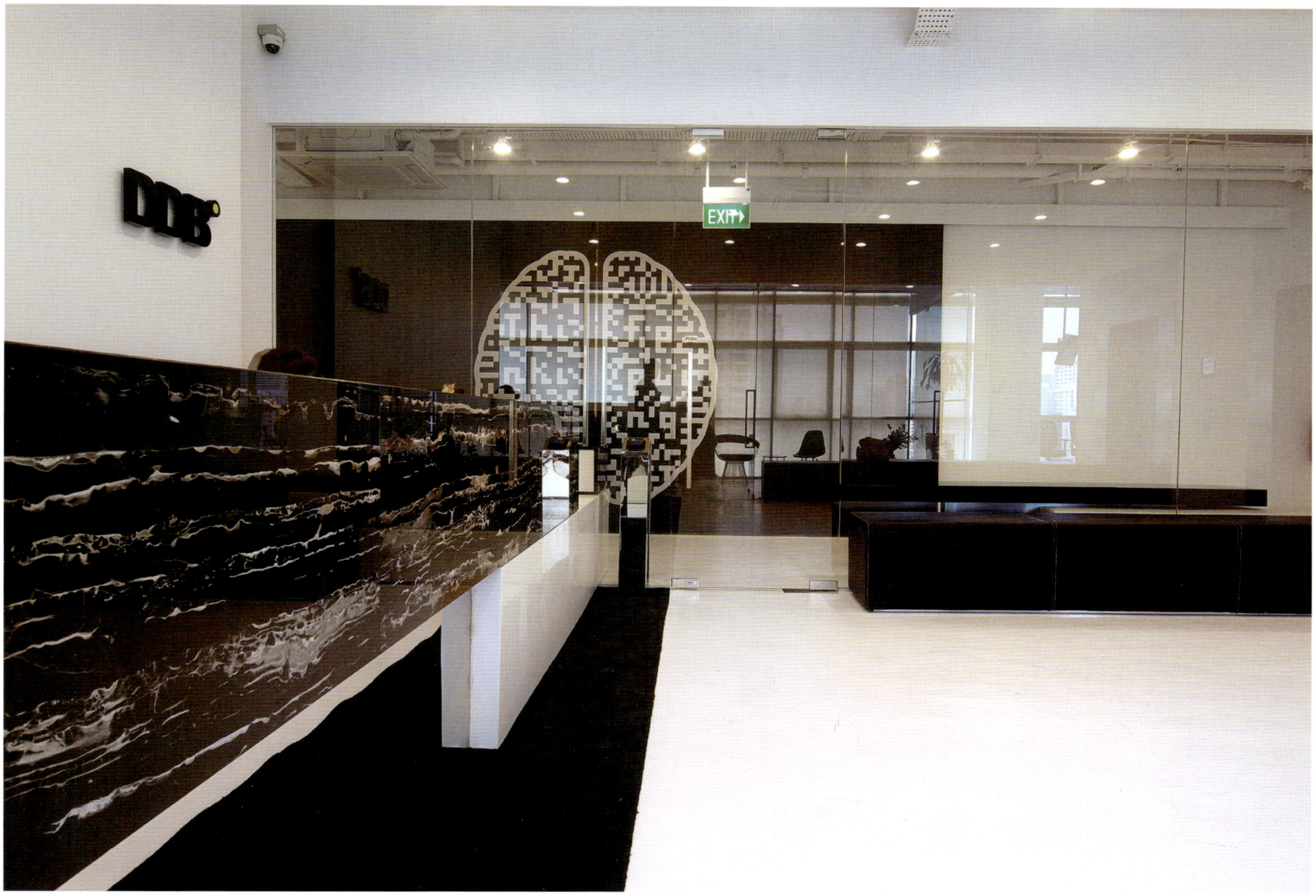
DDB
EXIT

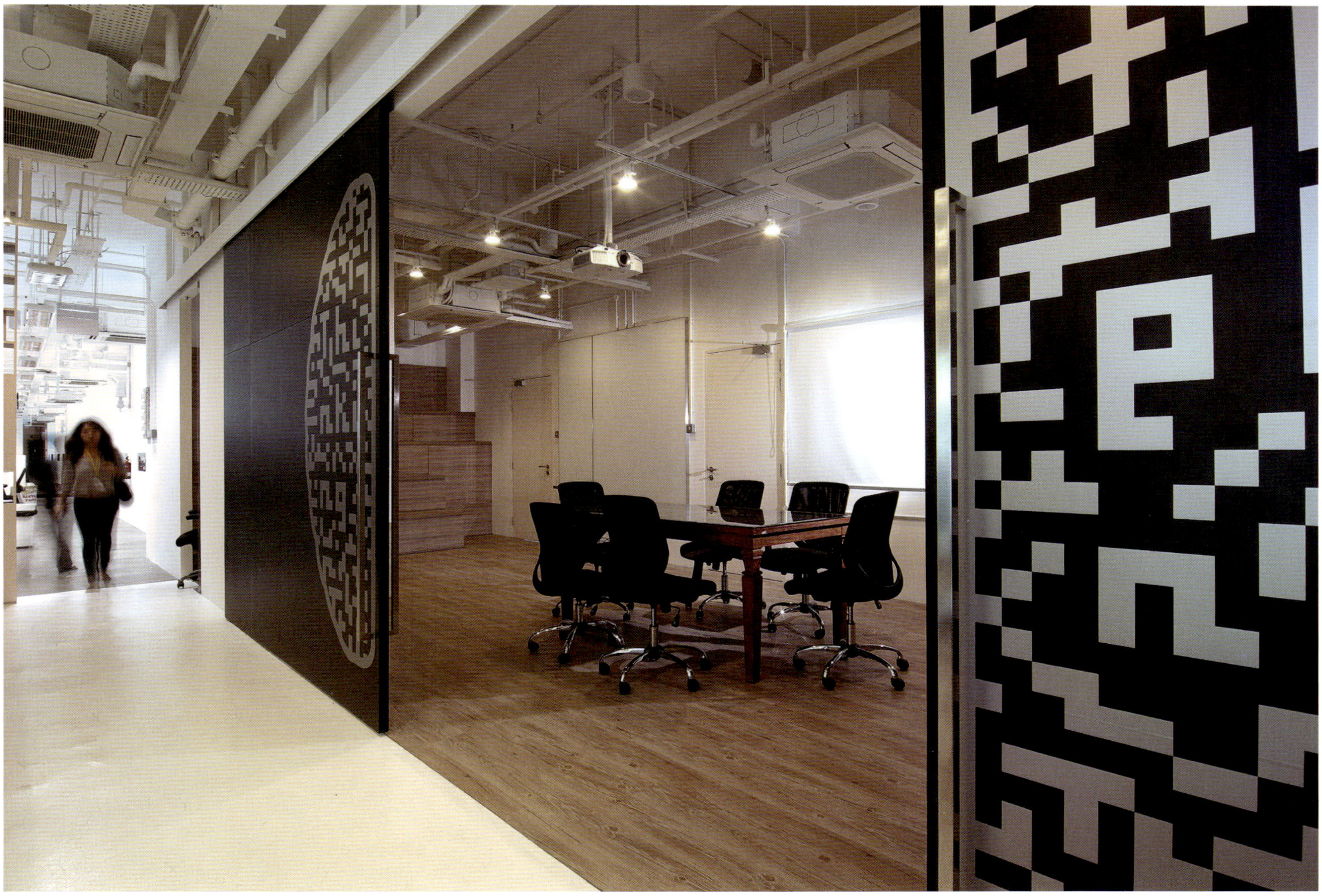

WORK*shop* _ISSUE.TWO

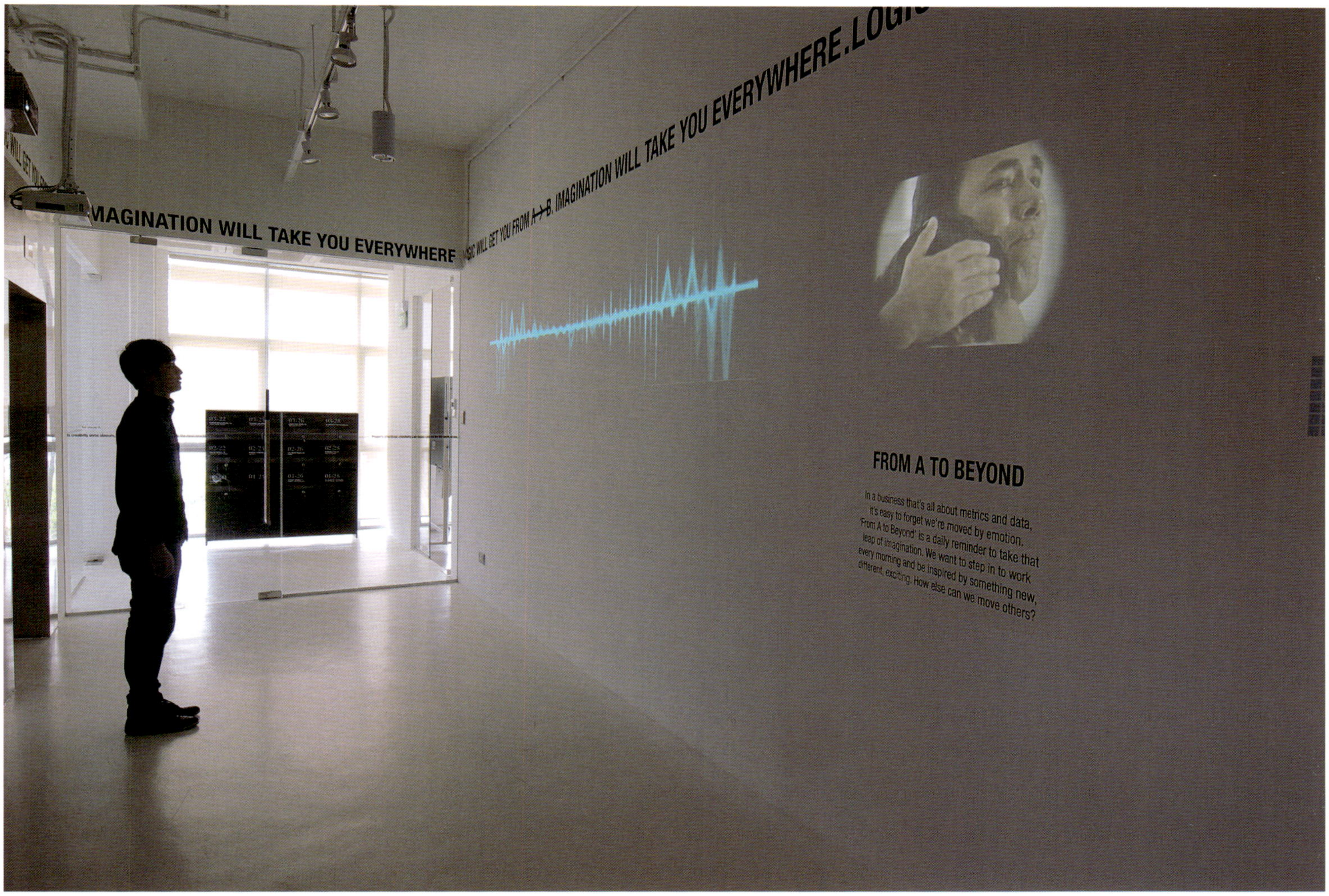

Territories are divided and screened by discussion and collaborative areas within the office. The aim was to enhance the current culture and to encourage a more lively work space, which will ultimately improve a better work flow.

The workspaces are designed as long communal tables supporting an open office environment. We design spaces that allow these communal areas to be part of the workspace.

Discussion bars are dispersed within team work areas. Even libraries and the staff pantry spanning across both floors were used as collaborative spaces for discussions.

Conference rooms were meant to be flexible and open where staff would frequently occupy for internal presentations and brain-storming sessions. Gym-like staggered benches were introduced for audiences to participate in these sessions.

Arrival lobbies on the other floors were treated with interactive projection of the collection of works that inspired DDB and a source of expression for the staff.

Recalls Shafie Latiff during the development stage of DDB's office design, 'To sculpt an environment for designers, it is essential that the spaces are open, flexible and inspiring where staff could express themselves freely just like a white open canvas.

DDB as a client has been a project which involves cross-disciplinary professions, allowing us as designers to push the forefront of art and design as a holistic approach.'

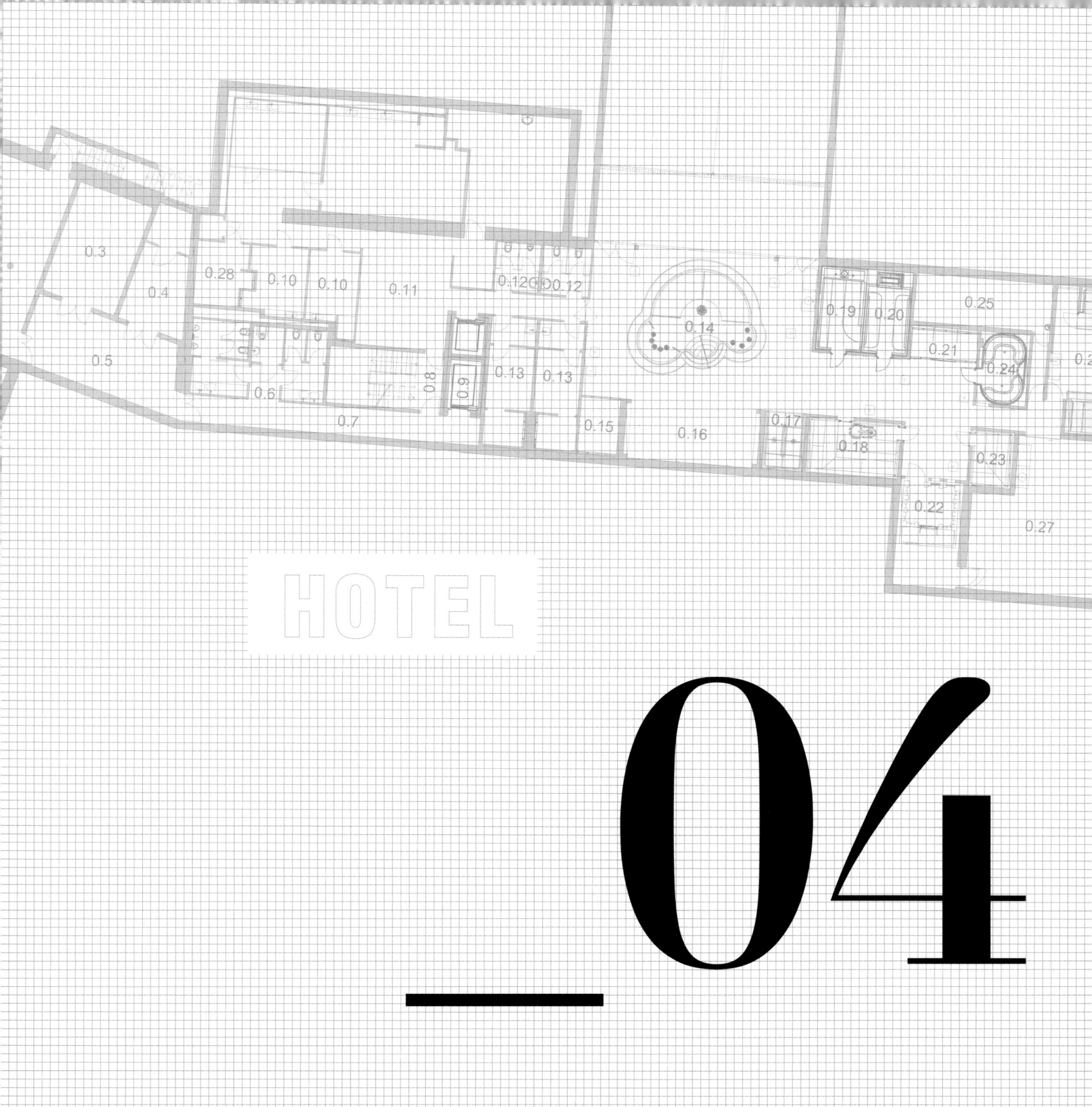

HOTEL

_04

HOTEL | Design_ LABOR13 | Photography_ Tomas Soucek, Martin Vomastek | Country_ Czech Republic | Client_ Miura Hotel

MIURA HOTEL

_LABOR13

The construction of the Miura Hotel is not only about the building. It is a sophisticated connection of architecture, design, graphic and art. It coexists here together and creates a complex with a clear concept and expression. Style of the hotel seems a little bit controversial. It is not a suggestive building that everybody likes. Its distinctive shapes follow the panorama of the surrounding hills and create a clear dominant aesthetics.

TERRACE

_ORIENTATION SYSTEM

_LOBBY

The hotel's building is located on a flat plain surrounded by hills of Beskydy Mountains and the lot fit directly into the famous Golf Course and became a part of the Course. In the neighborhood of the hotel there are few houses without any uniform architectural style. We decided to use original concept in this area. New designed building purposely breaks local scale and works as a new view point of the country.
From the beginning the hotel seems as a spaceship from another world. Actually this new element perfectly fits into the country. The idea of the concept is more developed in the sculptures (cube people), art and graphics. This unique world gives you a space for your imagination. You can find here unexpected elements and it's only up to you if you can find here your own story. There are masterpieces from the best Czech and international artists (Andy Warhol, David erný, Herny Moore Tony Cragge, John Armleder, Damien Hirst, Luca Pancrazzi, Petr Pastrňák).

_CONCRETE **RAMP**

_NIGHT **BAR**

The shape of the building reflects the subject. In the geometric centre there is a functional centre of the hotel – main entrance, lobby and a restaurant. These spaces are smoothly connected by concrete ramp and are opened to exterior by huge glass sheets. Location of the restaurant in the 2nd floor and glass walls offer panoramic views on the surrounding hills. To the west and east from the central part of the building there are hotel's wing parts. All rooms are opened to the south with the main view on the Golf Course. In the basement of the hotel there are located conference rooms and spa / wellness. Spa complex is naturally illuminated by glass wall. You can enjoy breathtaking views from outside terrace or directly from the whirlpool, but there is still kept the privacy of the spa. Dynamic castellated shape of the hotel is designed not to create a barrier in the country. This idea was supported by lifting the part of the hotel on piles so there is a view throughout the hotel and the design is more relieved.

South part of the object has the unique expression with random vertical windows of the rooms. Some of the windows have loggia with sharp colour. North façade of the hotel is structured with various large windows which are not follow the floors order. Colour scheme is minimalistic – shades of grey, black and white with contrast of fuchsia / magenta, which works as contrast colour to neutral shade.

There are used mostly natural materials – concrete, glass, stone, plate iron. Façade of the hotel is furred by coloured cembonit sheets.

_PLAN

LEGENDA MÍSTNOSTÍ

ozn.	místnost	ozn.	místnost
0.1	technické zázemí	0.15	sprchy
0.2	víceúčelový sál	0.16	odpočívárna
0.3	jednací místnost	0.17	kneipp
0.4	technické zázemí	0.18	sauna
0.5	předsálí	0.19	herbal bath
0.6	WC	0.20	sea climate
0.7	chodba	0.21	sprchy
0.8	schodiště	0.22	sněžná kabina
0.9	výtah	0.23	vario sauna
0.10	masáže	0.24	pára
0.11	recepce wellness	0.25	technické zázemí wellness
0.12	WC	0.26	salónek
0.13	šatny	0.27	darkroom
0.14	vířivka	0.28	sanotherm

WORK*shop* _ISSUE.TWO

THE CLUB

_Ministry of Design

| HOTEL | **Design**_ Ministry of Design | **Photography**_ CI&A Photography | **Country**_ Singapore | **Client**_ Harry's Hospitality Pte Ltd |

_SKY BAR **INDOOR**

_SKY BAR **OUTDOOR**

The Club is Ministry of Design's latest high design boutique hotel offering in the urban chic Club Street conservation area with 22 distinctly unique rooms, a rooftop sky bar with alfresco deck and a destination F&B venue with a tapas bar on the ground floor. Conceptualizing The Club's branding, MOD has orchestrated a unified design vision to all related collateral, signage and spatial environments. Targeted at the design and lifestyle savvy global nomad, the Club's blend of sophisticated and comfortable design is at once distinctly local as it is cutting edge global.

Colin Seah, Design Director says, 'Searching to ground the hotel in the context of Singapore as well as the historically rich conservation area of Club Street and Ann Siang Hill, we drew its inspiration from two sources.

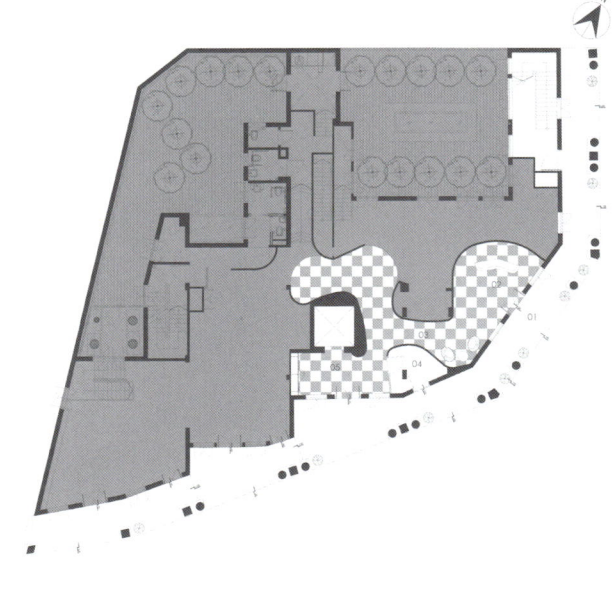

01 HOTEL ENTRANCE
02 RECEPTION
03 HALLWAY
04 LUGGAGE ROOM
05 LIFT LOBBY / RAFFLES STATUE

_GROUND FLOOR **PLAN**

_STATUE OF RAFFLES

'The first is Singapore's colonial past, which we have made modern tongue-in-cheek references to through art installation like features such as an larger-than-life statue of Raffles with his head in the clouds as well as through some key furniture pieces and artifacts.

'The second inspiration was drawn from the area's popularity as a remittance centre for turn of the century Chinese immigrants where hard earned money and wistful letters were sent back to the homeland. We have taken the memories of these exchanges and created features that hint of this legacy in the rooms of The Club, where the modern day nomad and the nomad of yesterday cross paths for a moment.'

All rooms combine traditional colonial design inspired elements together with sleek modern detailing, attitude and creature comforts – creating a colonial chic aesthetic. Unique layouts together with tailored artwork in each room make each of the 22 rooms distinct. MOD designed the artwork and famed local artist Wynlyn Tan implemented them in the hotel.

Guests have the option of checking in at the ground level lobby or at the panoramic roof top Sky Bar, overlooking the Club Street conservation area and CBD. F&B areas designed by Jane Yeo include Lobby Lounge, Tapas Bar, and two private function rooms.

_05

HEALTH/BEAUTY

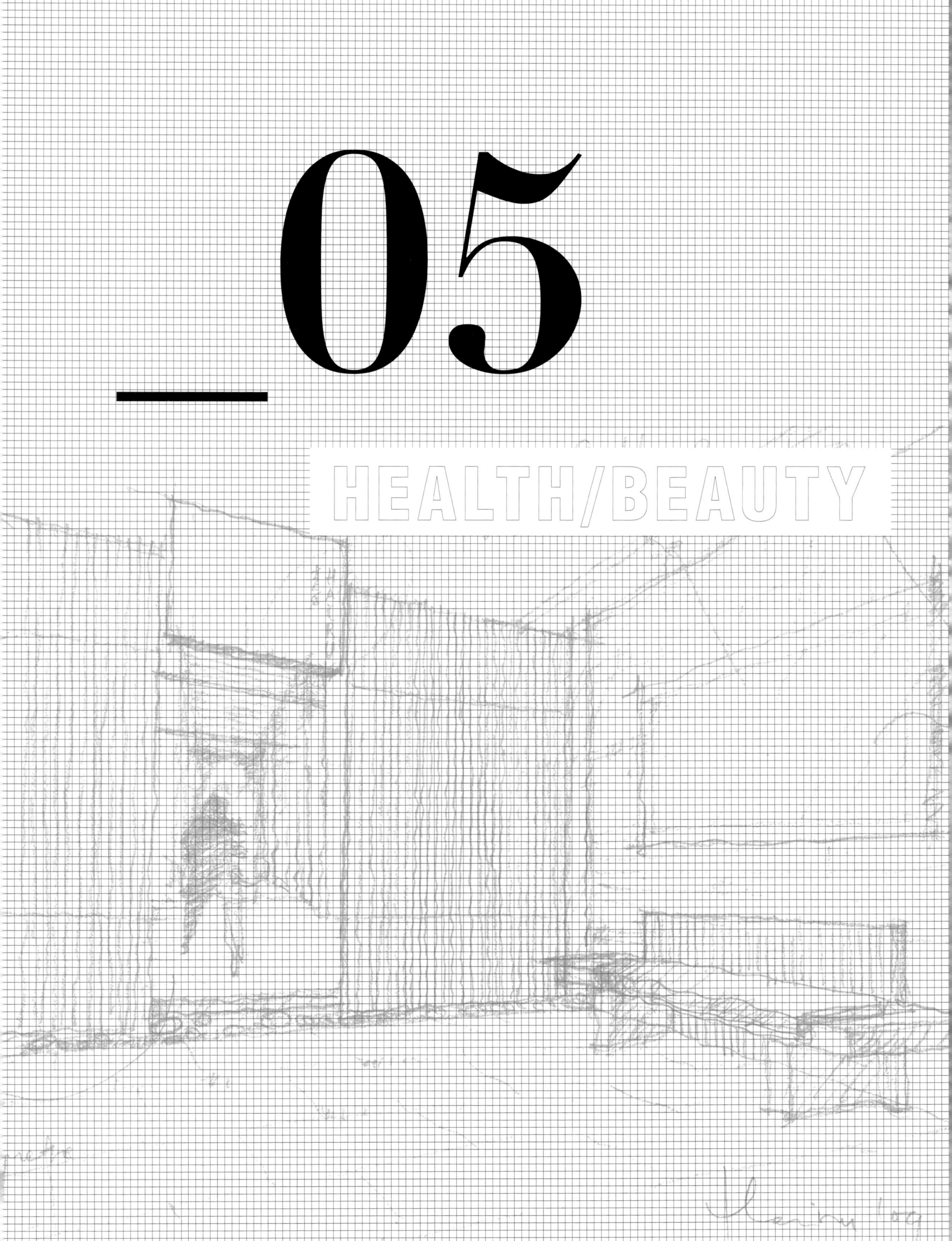

HAIRU HAIR TREATMENT

_Chrystalline Artchitect

'Creating a piece of sanctuary where you can relax with difference ambiance, the character of natural material is profound in this project.'
Started with modular design and built through generous clean and simple detail, performed in natural material scheme and colour, brought you to the difference experience of leisure. It may look like a salon, but Hairu is engaged in hair health-care, such as treatment for losing hair, hair spa and massage.
The vocal point is the double faced rough bali-an stone which is also a transition door to the wash area. The ribbon mirror on the both side walls is carried through the view of all the space with the private sight just for the client without having eye contact with the therapist.
The soft material, vitrage, is acting as a divider which could be moved to get along with the other. And the hard material, cotton, is as a permanent divider from the main corridor.

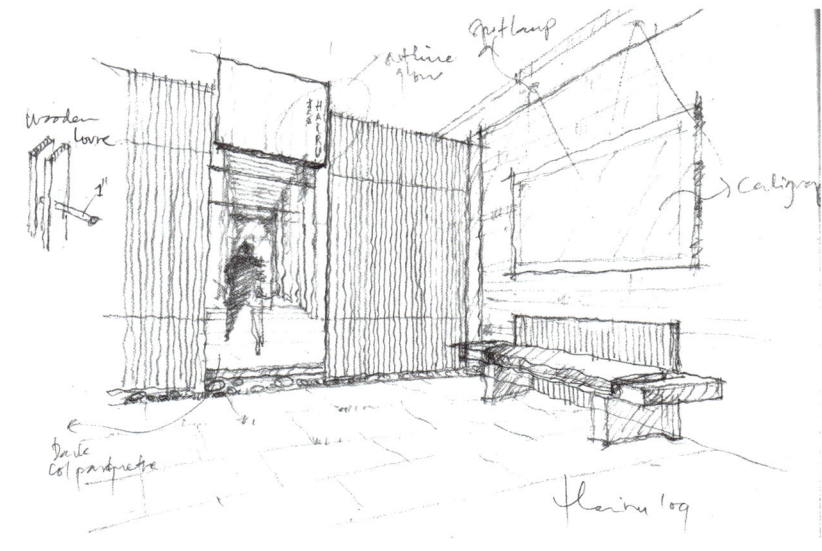

_ENTRANCE SKETCH

HEALTH / BEAUTY | Design_ Chrystalline Artchitect | Photography_ William Sebastian | Country_ Indonesia | Client_ Private

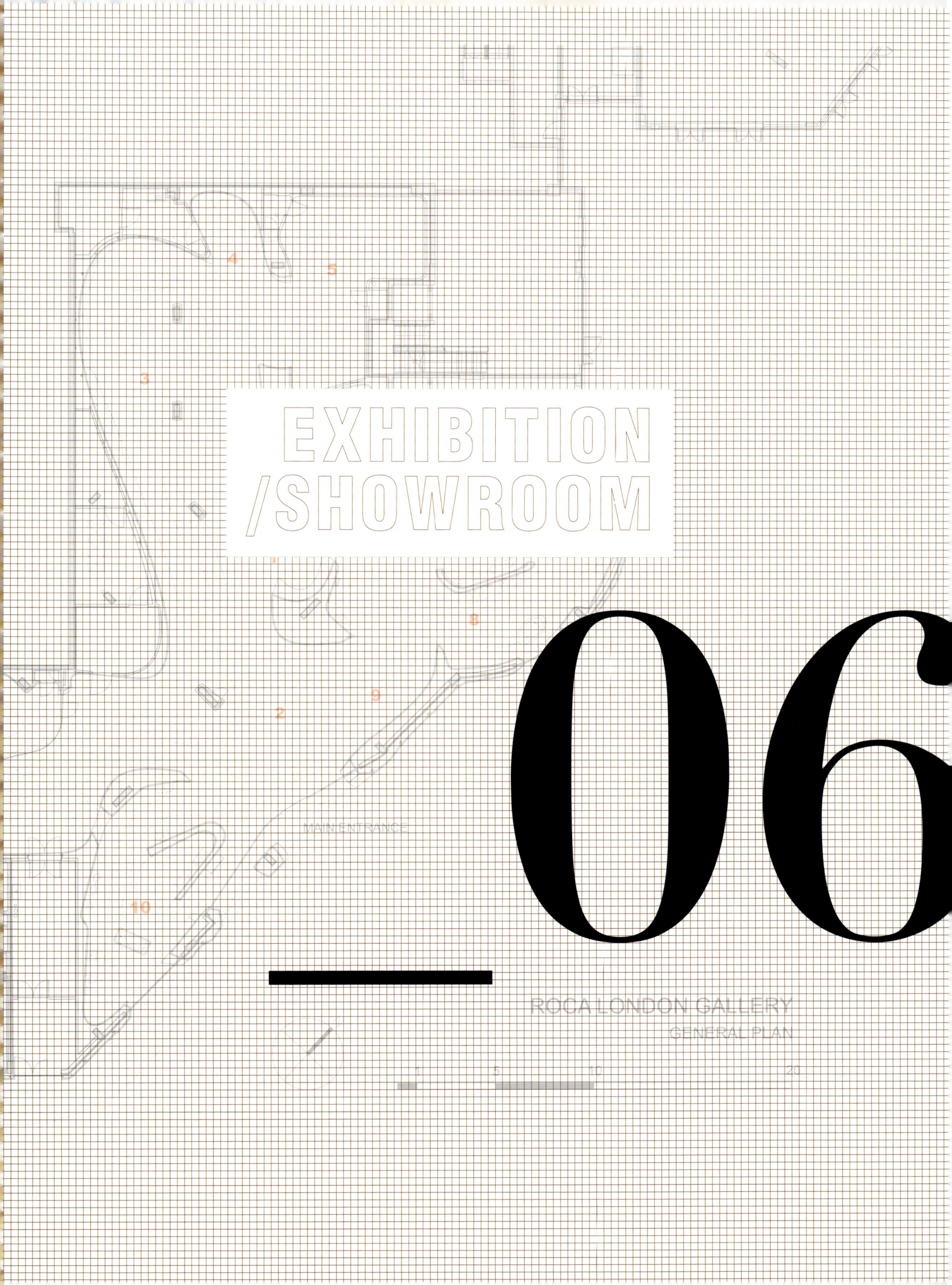

EXHIBITION /SHOWROOM

06

ROCA LONDON GALLERY
GENERAL PLAN

BRUNNER SALONE INTERNAZIONALE DEL MOBILE 2011

_Ippolito Fleitz Group

A brand new product highlight takes centre stage on the Brunner exhibition stand at the Milan Furniture Fair: twin, a monobloc plastic chair. Brunner enters a new price segment with the introduction of this exceedingly lightweight and reasonably priced chair. The company aims to establish a presence in the design product sector as well as attracting architects as a disseminating force.

A whole world geared around the slogan See / Reflect / Act has been created, in which the furniture is not simply a display item, but a vibrant protagonist in a scene that resembles an art installation. The stand is accompanied by a campaign consisting of invitations, giveaways and carrier bags.

In the form of an elongated rectangle accessible from the narrow front end, the stand dissolves into a dynamic space thanks to curving walls along the longitudinal sides and rear end, which pull the visitor inside the cavernous interior. To your right hand-side when entering the stand, a loose grouping of different twin models gives the upbeat. From the midst of this group, a single chair is elevated to eye level. This forms the starting point for a dynamic whirl of several dozen chairs suspended in a cloud above the visitors' heads. These execute a backwards somersault wherein each adjacent section is rotated through 45 degrees. The resulting ensemble enables the visitor to view the chair from every conceivable angle. The swarm of chairs returns to its starting position at the opposite wall, where it is pulled back down to floor level. Here the entire spectrum of available models is presented and the visitor is invited to sit and test the chairs.

The walls of the space are completely panelled with mirrored polystyrene shingles. The panoramic effect of the mirrors adds additional dramatic impact to the scene. The mirrors greatly increase the already large number of chairs and their colour palette appears to spill down the walls, mixing with the reflections of the visitors. Their honeycomb structure deconstructs reflections: individual forms are no longer recognisable and everything melts into a pixellated burst of colour. The wall shingles also hark back to the company's Black Forest origins, conjuring up associations of tradition and high quality. Individual shingles printed with product information are dispersed across the chairs and areas of artificial grass. Visitors can thus take home their very own piece of the exhibition stand as a giveaway.

The stand's organic layout is underscored by two zones of white artificial grass along the longitudinal walls as well as an island in the centre, demarcating the lounge that is furnished with other Brunner products. A counter positioned in front of the curved long wall, behind which kitchen and storeroom are cleverly concealed, serves as an additional communication zone.

The new twin chair celebrates a spectacular premiere on this exhibition stand. The whole stand becomes a stage, wholly encapsulating the visitor and drawing him into an intense spatial experience.

_BRUNNER – SALONE INTERNAZIONALE DEL MOBILE 2011   _EXHIBITION/**SHOWROOM**

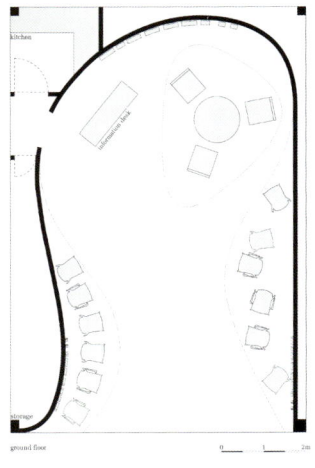

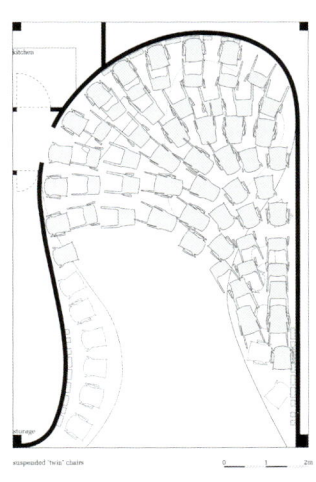

_FLOOR **PLAN**

Client_ Brunner GmbH

THE DESIGN BAR

_Jonas Wagell Design & Architecture

Trade fairs are temporary constructions quickly built to be torn down shortly after. Instead of creating an installation which attempts to be more than this, the Design Bar of year 2010 aims to embrace the temporary by creating a space which is influenced by stage design and graphics rather than polished architecture.
In year 2010 the Design Bar was commissioned to Swedish architect and designer Jonas Wagell who – by small means and measures – created an expressive space with strong character.
There were also a VIP lounge which shared the same conceptual idea, but with different look and feel.

The bar area served light foods and provided a surrounding where you can sit back and enjoy a drink or coffee, while the secluded VIP area would offer an undisturbed seating for meeting and chats.
The conceptual theme for the Design Bar and the VIP Lounge was Forest and Industry – a tribute to raw materials, craftsmanship and refinement, which constitute the backbone of the furniture industry.
The Design Bar was furnished with new products by Jonas Wagell – the 'Montmartre' furniture set from Mitab, 'Mr Gardner' outdoor easy chairs from Berga Form and the 'Cage' steel baskets and the 'Odd' family of pendant lamps, bowls and vases from Hello Industry.

_THE **DESIGN BAR** _EXHIBITION/**SHOWROOM** _161

Client_ Stockholm Furniture Fair

GE HEALTHYMAGINATION SHOWCASE

_Urban A&O, Thinc Design, Local Projects

Conceived as a 'gravity-free space of creative imagination', the GE Healthymagination Showcase developed into an experience organised around three 'pods' and a central gathering space. Urban A&O, Thinc Design and Local Projects partnered to design and produce the event.

Urban A&O developed the sinuous, organic overhead pod structures and their corresponding ramped floors in CATIA, which allowed them to rapidly generate the extremely complex, flowing forms that became the signature of the project. The overhead structures were developed as three-dimensional reticulated surfaces: networks of division, marking and construction that form a network of lines and surfaces inspired by the intricate structures of the human body. What begins as two-dimensional patterning is drawn into three dimensions, generating suspended forms comprised of sinuously-shaped, volumetric ribs with curvilinear voids between them. The GE Healthymagination Showcase in New York debuted GE's encompassing health care initiative for its major customers and constituents. Occupying a prominent storefront location at 44th Street and Fifth Avenue, it offered a range or programming to physicians, hospital administrators, stock analysts, press, GE staff, government officials and the general public.

CENTRAL HOSTING **SPACE**

THE ORANGE CUBE AND RBC

_Jakob + MacFarlane Architects

The ambition of the urban planning project for the old harbor zone, developed by VNF (Voies Naviguables de France) in partnership with Caisse des Dépôts and Sem Lyon Confluence, was to reinvest the docks of Lyon on the river side and its industrial patrimony, bringing together architecture and a cultural and commercial program.

These docks, initially made of warehouses (la Sucrière, les Douanes, les Salins, la Capitainerie), cranes, functional elements bound to the river and its flow, mutate into a territory of experimentation in order to create a new landscape that is articulated towards the river and the surrounding hills.

The project is designed as a simple orthogonal cube into which a giant hole is carved, responding to necessities of light, air movement and views. This hole creates a void, piercing the building horizontally from the river side inwards and upwards through the roof terrace.

The cube, next to the existing hall (the Salins building, made from three archs) highlights its autonomy. It is designed on a regular framework (29x33m) made of concrete pillars on 5 levels. A light façade, with seemingly random openings is completed by another façade, pierced with pixilated patterns that accompany the movement of the river. The orange colour refers to lead paint, an industrial colour often used for harbor zones.

In order to create the void, Jakob + MacFarlane worked with a series of volumetric perturbations, linked to the subtraction of three 'conic' volumes disposed on three levels: the angle of the façade, the roof and the level of the entry. These perturbations generate spaces and relations between the building, its users, the site and the light supply, inside a common office program.

The first perturbation is based on direct visual relation with the arched structure of the hall, its proximity and its buttress form. It allows to connect the two architectural elements and to create new space on a double height, protected inside the building.

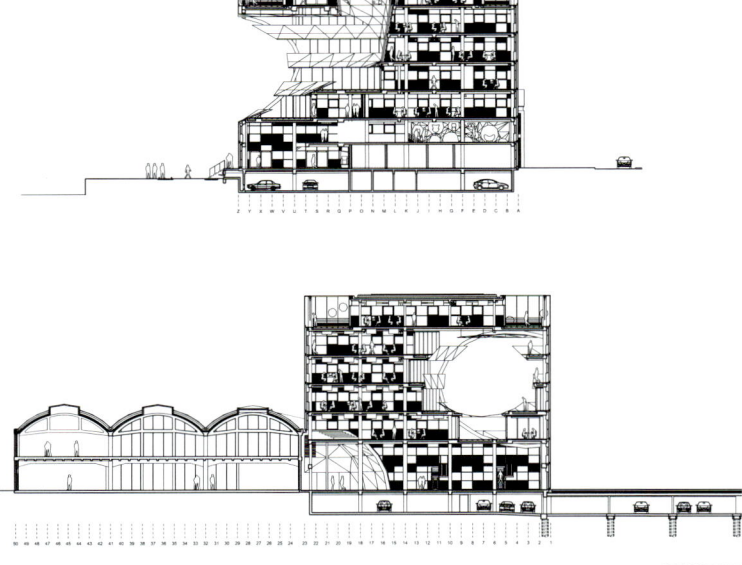

_SECTION

_THE ORANGE CUBE AND **RBC** EXHIBITION/**SHOWROOM**

Client_ Rhône Saône Développement

The second, obviously an elliptic one, breaks the structural regularity of the pole-girder structure on four levels at the level of the façade corner that gives on the river side. This perforation, result of the encounter of two curves, establishes a diagonal relation towards the angle. It generates a huge atrium in the depth of the volume, surrounded by a series of corridors connected to the office platforms. The plan of the façade is hence shifted towards the interior, constructing a new relation to light and view, from both interior and exterior. This creates an extremely dynamic relation with the building that changes geometry according to the position of the spectator. The tertiary platforms benefit from light and views at different levels with balconies that are accessible from each level. Each platform enjoys a new sort of conviviality through the access on the balconies and its views, creating spaces for encounter and informal exchanges. The research for transparency and optimal light transmission on the platforms contributes to make the working spaces more elegant and light.

The last floor has a big terrace in the background from which one can admire the whole panoramic view on Lyon, la Fourvière and Lyon-Confluence.

The project is part of the approach for sustainable development and respects the following principles: optimization of the façade conception allowing to reconcile thermal performance and visual comfort with an Ubat < 0.7 W / m2 K and a daylight factor of 2% for almost the total number of offices, a thermo frigorific production through heat pumps on the water level and the replacement of new hygienic air with recuperation of high efficient calories of the extracted air.

The building is connected to future huge floating terraces connected to the banks of the river / quays.

Showroom Concept

This project was about bringing together a showroom dedicated to the world of design objects inside the architecture of an existing building: the Orange Cube. The intention was to bring the worlds of architecture, design and the uniqueness of the site in Lyon together into one experience.

Jakob + MacFarlane decided to take the language of the Cube, which is based on the fluid movement of the River Saône and in a sense project this movement inside the space of the showroom. Thus imagining the space as an extrapolation of the façade, a virtual three-dimensional river or volume containing a long porous wall whose 60 'alvéoles' are filled with furniture. This wall wraps around the space of the showroom forming an 'L'. The spectator moves from the spectacular entry wall towards more intimate spaces on the river side. Each 'alvéole' is unique in size and form allowing thus an intimate and private view of each design piece.

The platforms on the floor, made from a series of kitset pants, imagined like islands, can become stages for different thematic presentations.

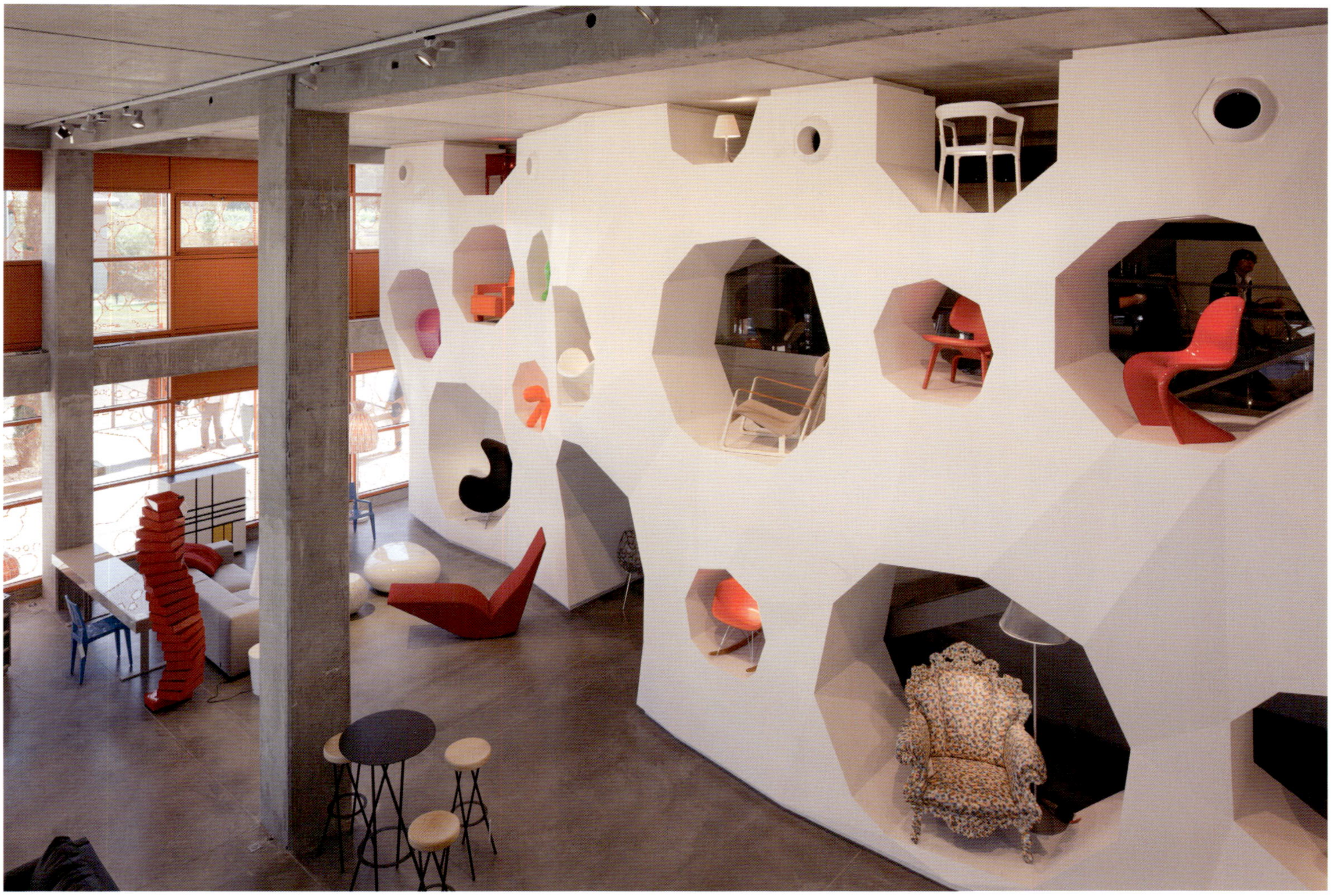

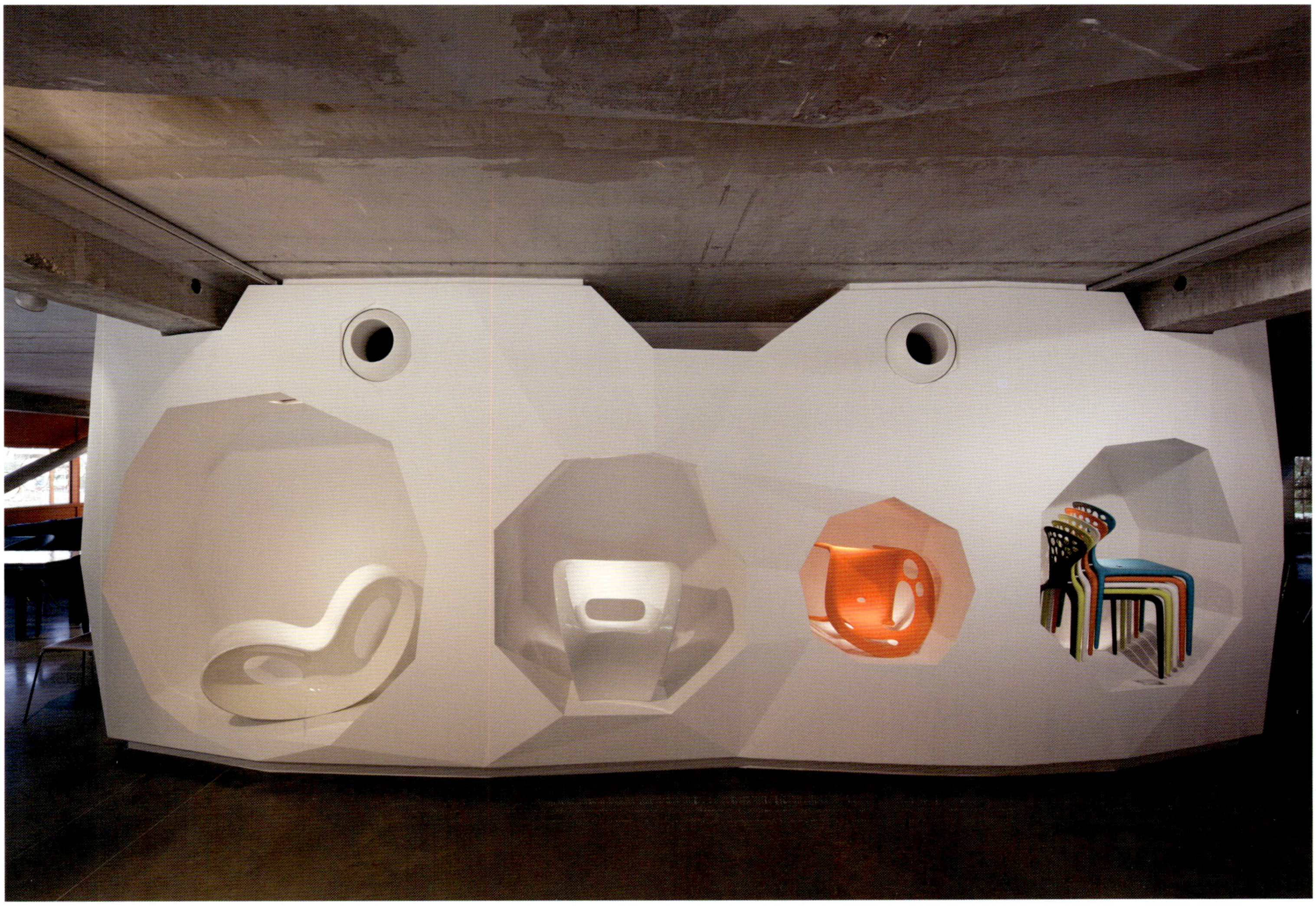

WORK _ISSUE.TWO

CHOPIN'S VISITING CARD

_Boris Kudlicka and WWAA

The exposition 'Chopin's visiting card' in The Krasiski Palace in Warsaw is a presentation of the biggest collection in the world of music manuscripts of Frederic Chopin which are in the archive of the Polish National Library. Next to the unusually rarely shown manuscripts (including 73 music pieces) Chopin's letters will be presented. The showpieces will be accompanied by interactive applications through touchscreens and headsets allowing to explore the sound level of the exhibition.

EXHIBITION / **SHOWROOM** **Design**_ Boris Kudlicka and WWAA **Photography**_ Aleksander Rutkowski and WWAA **Country**_ Poland

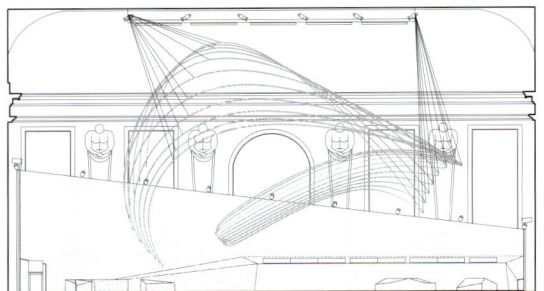

_SECTION

The leitmotiv of the exhibition are the sources of inspiration, creation process and the variety of forms used by Frederic Chopin. The aim of the exhibition design was to transfer the visitors into other reality – abstract world of the composer, in which his music and the sound of piano is a guide. Abstract spatial form wondering through the successive rooms of the palace is leading the visitors not only through the exhibition but first of all through the creation process of Frederic Chopin. Initially chaotic, tangled singular ribbons symbolizing improvisation and creation of first ideas of melodies unite and become smoother to create finally a perfect form filling the space of the upper exhibition room. Thin stripes of banded plywood and laborious process of deforming them can be associated with a work of creating both musical instruments and musical pieces. The background for the spatial form made of the plywood stripes is the dark floor folded to become partitions with printed texts and covers for the showcases in which exhibition pieces are placed. Vast seating in the same colour and form of the showcases serve as a place to sink into the Chopin's music.

WORK*shop* _ISSUE.TWO

EXHIBITION IN THE POLAND PAVILION FOR EXPO 2010 IN SHANGHAI

_WWAA + Boris Kudlicka

General Assumptions
The interior and the exhibition design of the Polish Pavilion for the Expo 2010 in Shanghai is a continuation of the architectural idea of the form of the building and the details of the facade. The aesthetic concept of the pavilion is brought inside and the folk cut-outs lead the visitors through the entrance into the main hall and then, while transforming into other patterns, are continuously guiding them through all the exhibition. The usage of the cut-out patterns has not only an aesthetic value, but also an educational function associated with the main theme of the EXPO: 'better city – better life'. The cut-outs changing from the folk forms into organic ones and finally into a city-map and industrial patterns are a metaphor of migration of people from countryside into cities. The story that the patterns are suppose to tell is the base for the presented images and films showing Poland through its history, culture, economy and everyday life. The design of the cut-outs goes with the presented on it contents changing along the visitors' route.

EXHIBITION / SHOWROOM | Design_ WWAA + Boris Kudlicka | Photography_ WWAA | Country_ Poland

Spatial Geometry

The main element defining the space in the pavilion is an internal wall with a complex geometry resembling a creased piece of paper. The wall determines the visiting route, divides the exhibition into zones and serves as a background for the multimedia presentations. The geometry of the wall is changing with its high varying from the 2.5m up to 7m. The main partition with the side walls, which have similar geometry, create a corridor leading the visitors. Some parts of the tunnel are covered with an openwork modular structure imitating greenery.

Client_ Polish Agency for Enterprise Development (PARP)

WORK*shop* _ISSUE.TWO

PRAGUE QUADRIENNALE 2011

_WWAA + Boris Kudlicka

A sketch is an initial phase of scenic design. It gives main impression on the project, shows the clue, indicates densities of the scenic design's plans. We use sketches as an emphasis of the theme for Prague Quadrennial of Performance Design and Space and make them three-dimensional.

Surfaces made of drafts, technical as well as freehand, divide the Veletrzni Palace into functional zones, establish spatial hierarchies and, what is most important, redraw the building as an exhibition space. Using the technique of drawing, we create surfaces of different density and also some geometric figures in space. That way it's possible to emphasise the artificiality of our spatial arrangement for PQ 2011 and its temporary character.

The surfaces of freehand sketches that make walls, screens, lounge zones or even furniture, are made of rubber cables differently braided or something that resembles it, according to their position in space. The line can be made of cable, expander or sticker. The PQ design is drawn with a continuous line that is changing its style due to the character of space it wants to evoke in the Veletrzni Palace. The line of the electric cable starts before the building, continues throughout the Palace just to end on the rooftop, at the PQ sky bar.

The atmosphere of PQ 2011 is visible from the start – the ticket pavilion is where the line starts, where it begins creating the exhibition space. Then the line goes through corridors' floor, walls and ceilings, sometimes in ordered way, sometimes it is absolutely spontaneous. The line leads us through successive zones with of curtains. In the exhibition area 'the line' is shy, and discretely flits through the middle of halls. The interference is delicate, almost invisible in the exhibition space laden with different forms, volumes and colours. On the contrary, so visually strong and obtrusive is the wired cloud in a high hall near the gallery entrance. It appeals heavy, but when observed from the balconies, surprisingly enlightens. In another zone of Veletrzni Palace 'the line' creates ordered structure under the ceiling that might be the smoke rising above lounge area. One of branches goes to the video

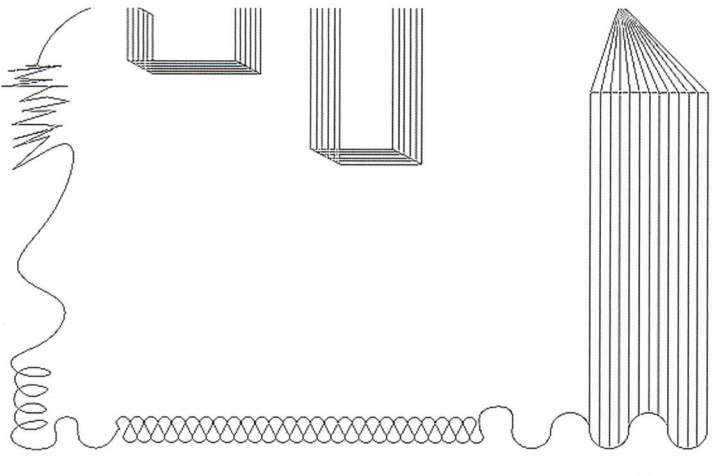

_SCHEME

EXHIBITION / SHOWROOM   Design_ WWAA + Boris Kudlicka   Photography_ WWAA + Jakub Certowicz   Country_ Poland

Client_ Prague Quadriennale 2011

WORK*shop* _ISSUE.TWO

lounge, where wrapes along walls make cocoon – audience. For the staircase we use black label to show how the line with a stairway motif modifies the staircase's volume. Up on the last floor, we astonish the audience with a folding shape of PQ auditorium. Wire lines overlap themselves, giving the curtain different densities from different perspectives. The line ends up in the terrace, wrapped around the barrels that serve as a sky bar and its seats. It intensifies once more in the sign 'love' to vanish in the sky.

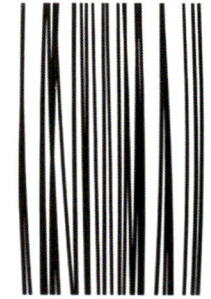

DECAMERON

_Studio MK27

The showroom of the Decameron furniture store is located on a rented site in the furniture commercial alley in São Paulo. To make the quick and economic construction viable, the project worked with the premise of a light occupation of the lot, basically done with industrial elements, which could easily be assembled.

The space was constructed through a mixed solution, with maritime transport containers and a specifically designed structure. Despite the spatial limitation imposed by the pre-determined dimension of the containers, the piece has impressive structural attributes that makes piling them possible. Two stories of containers form tunnels where products are displayed side by side.

The ample span, necessary to show furniture in relation with each other, is constructed by a metallic structure. This space is closed, in front and in back, by double-height metal casements with alveolar polycarbonate. At the back of the lot, there is a patio filled with trees and a pebbled-ground. When both doors are simultaneously opened, the whole store becomes integrated with its urban context. At rush stressful hours, by opening only the back doors, the store becomes self-absorbed, ruled by the presence of the inner-garden. On the back of the site is the office, closed by a glass wall that enables the designers to take part on the sales life. Two edges of the design process in contact through the inner patio as other opposing strengths also meet at this small project: the intensity of the urban life and a small nature retreat, the power of the containers and the lightness of the metallic structure and finally, the linearity of the tunnels and the cubic volume.

_SECTION

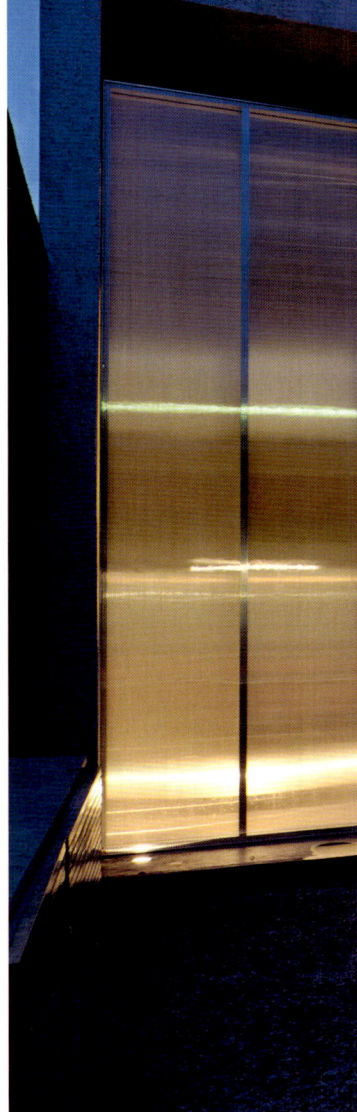

| EXHIBITION / SHOWROOM | Design_ Studio MK27 | Photography_ Pedro Vannucchi | Country_ Brazil | Client_ Decameron furniture store |

WORK*shop* _ISSUE.TWO

On the back of the site is the office, closed by a glass wall that enables the designers to take part on the sales life.

DOCKS EN SEINE

_Jakob + MacFarlane Architects

The docks of Paris is a long, thin building built in concrete at the turn of the last century. It was a depot for goods brought up the Seine by barge, which were deposited, and then transferred to dray or train.

The city of Paris launched a competition to create a new cultural program and building on this site. Whether or not to keep the existing concrete structure was a choice left to the participants. Jakob+MacFarlane opted to retain the existing structure and use it to form and influence the new project.

The existing structure was built in 1907 as an industrial warehouse facility for the Port of Paris and was the first reinforced concrete building in Paris. The three-storey structure was conceived as a series of 4 pavilions, each with one 10m wide bay and four 7.5m wide bays. On the level corresponding to the Quai Austerlitz, the 10m bay is accessible from the street with the other bays roughly 1.25m higher, facilitating the storing and loading of materials for transport.

The concept of the new project is known as a 'Plug-Over'. The idea was to create a new external skin that is inspired by the flux of the Seine and the promenades along the sides of the river banks. An arborescent generating method is used to create a new system from the existing system, which is, 'growing' the new building from the old as new branches growing on a tree. This skin is created principally from a glass exterior skin, steel structure, wood decking and grassed, faceted roofscape. The 'Plug-Over' operates not only as a way of exploiting the maximum building envelope but enables a continuous public path to move up through the building from the lowest level alongside the Seine to the roof deck and back down.

The programme is a rich mix centred on the themes of design and fashion, including exhibition spaces, the French Fashion Institute (IFM), music producers, bookshops, cafés, and a restaurant.

Exhibition / Showroom | Design_ Jakob + MacFarlane Architects | Photography_ Nicolas Borel | Country_ France

WORK*shop* _ISSUE.TWO

ZAHA HADID: FORM IN MOTION

_Zaha Hadid

EXHIBITION / SHOWROOM   Design_ Zaha Hadid   Photography_ Paul Warchol   Country_ UK

Zaha Hadid, one of the most innovative architects of the twenty-first century and the first woman to receive the renowned Pritzker Architecture Prize in 2004, has advanced the language of contemporary architecture and design, exploring complex fluid geometries and using cutting-edge digital design and fabrication technologies.
Born in Baghdad, Iraq, in 1950, Zaha Hadid studied in Lebanon, Switzerland, and in England. Today, Hadid, a British citizen, founding director of Zaha Hadid Architects, is based in London and works on projects throughout the world. Recently completed projects include the Guangzhou Opera House in China; MAXXI: National Museum of 21st Century Art in Rome; the Riverside Museum of Transport in Glasgow and the Aquatics Centre for the London 2012 Olympic Games.

For *Zaha Hadid: Form in Motion* (September 17, 2011 to March 25, 2012) – the first in the United States to feature her product designs Hadid – has created a sculptural environment for a selection of furniture, decorative art, jewelry, and footwear that she has designed in recent years as well as the prototype for her *Z-Car I* (2005).

Combining architecture and design, *Zaha Hadid: Form in Motion* displays an all-encompassing environment of an undulating structure of finished polystyrene with vinyl graphics based on curvilinear geometries. Exploiting a formal language of fluid movement, Hadid's exhibition design emphasises the continuous nature of her work, reinventing the balance between objects and space in an interior landscape, and how the fields of architecture, urbanism, and design are closely interrelated in her practice.

Sleekly curving sofas, tables, and lounge chairs made of materials ranging from wood, steel and aluminium to polyurethane inhabit the gallery, while jewellery, shoes, and tableware are installed together in small groups along a rippling wall representing the wide variety of new and unusual shapes Hadid has introduced into the language of design. The *Mesa* Table is supported by branching, lofted connectors, more void than solid, while a table made of polished aluminium appears to hover close to the floor supported only by the same invisible forces that generate the craters on its surface. The striated video wall, sinuous floor and wall graphics transform the gallery and its contents into a singular, fluid, dynamic composition.

Form in Motion displays an all-encompassing environment of an undulating structure of finished polystyrene with vinyl graphics based on curvilinear geometries.

Some works are disguised as micro-architecture, such as the *Coffee & Tea Set* (1997), nearly unidentifiable as a set of containers for tea, coffee, milk, and sugar. Others, including *WMF Flatware* and *Crevasse Vases*, are more transparent in function. Among the highlights are a collection of Swarovski crystal-encrusted necklaces and bracelets, and spiralling, strappy shoes made for Lacoste and Melissa. Hadid's three-wheeled *Z-Car I*, an aerodynamic prototype created of high-density foam that echoes several of her sculptural forms, will be on view in the Perelman Atrium.

On November 19, 2011, Zaha Hadid was honored with the Design Excellence Award given by Collab, a group of design professionals and enthusiasts who support the Museum's modern and contemporary design collection. Collab's Student Design Competition (offered since 1993) challenges area college students studying architecture and industrial design to be inspired by themes closely associated with the Design Excellence Award winner and the corresponding exhibition.

WORKshop _ISSUE.TWO

ZAHA HADID, UNE ARCHITECTURE IN MOBILE ART PAVILION

_Zaha Hadid

EXHIBITION / SHOWROOM **Design_** Zaha Hadid **Photography_** Francois Lacour **Country_** UK **Client_** Institut du monde arabe

On 28 April, 2011, the exhibition which showcased a selection of work by the 2004 Pritzker Prize laureate Zaha Hadid, designed by herself, inaugurated the Mobile Art Pavilion, a new arts venue installed in front of the Institut du monde arabe.

Mobile Art Pavilion

'Zaha Hadid' was the first exhibition held inside the Mobile Art Pavilion since the installation of the pavilion in front of the Institut du monde arabe. CHANEL donated the pavilion to the Institut du monde arabe at the beginning of 2011. It had previously travelled to Hong Kong, Tokyo and New York since created by Iraqi born British architect Zaha Hadid for CHANEL in 2007. Now it has a permanent location at the IMA, where it is used to host exhibitions in line with the centre's policy of showcasing talent from Arab countries.

Zaha Hadid Architects' recent explorations of natural organizational systems have generated the fluidity evident in the Mobile Art Pavilion. The Mobile Art Pavilion's organic form has evolved from the spiraling shapes found in nature. This system of organization and growth offers an appropriate expansion towards its circumference, giving the Pavilion generous public areas at its entrance with a 125m² terrace. The Pavilion follows the parametric distortion of a torus. In its purest geometric shape, the circular torus is the most fundamental diagram of an exhibition space. The distortion evident in the Pavilion creates a constant variety of exhibition spaces around its circumference, whilst at its centre, a large 60m² courtyard with natural lighting provides an area for visitors to meet and reflect on the exhibition. This arrangement also allows visitors to see each other moving through the space and interacting with the exhibition. In this way, the architecture facilitates the viewing of art as a collective experience.

The organic fibre reinforced plastic shell of the Mobile Art Pavilion is created with a succession of reducing arched steel segments. As the Pavilion has traveled over three continents, this segmentation also gives an appropriate system of partitioning – allowing the Pavilion to be easily transported in separate, manageable elements. Each structural element is not wider than 2.25m. The partitioning seams become a strong formal feature of the exterior façade cladding, whilst these seams also create a spatial rhythm of perspective views within the interior exhibition spaces. The Mobile Art Pavilion which has been conceived through a system of natural organization, is also shaped by the functional considerations of the exhibition. However, these further determinations remain secondary and precariously dependent on the overriding formal language of the Pavilion. An enigmatic strangeness has evolved between the Pavilion's organic system of logic and these functional adaptations – arousing the visitor's curiosity even further.

In creating the Mobile Art Pavilion, Zaha Hadid has developed the fluid geometries of natural systems into a continuum of fluent and dynamic space – where oppositions between exterior and interior, light and dark, natural and artificial landscapes are synthesised. Lines of energy converge within the Pavilion, constantly redefining the quality of each exhibition space whilst guiding movement through the exhibition.

Zaha Hadid, une Architecture (29 April - 30 October 2011)

The pavilion's interior in its refurbished status was enriched by fluid forms that structure the torus-like exhibition loop and therefore differentiate the visitors' tour. The aim was to utilise the existing quality of the sculpted internal walls and space divisions to enhance the visitors' experience and provide rhythmical spatial sequences for a sensually differentiated walk-through.

A parametrically derived special structure was unfolded within the pavilion to exhibit, display and organise thematic hotspots. Media interventions such as projected animations were placed at strategic points within the exhibition design. Particular points of interest were generated with highlights from the body of work of Zaha Hadid in the form of architectural models, silver paintings and wall reliefs.

The exhibition installation was a scenographic background and canvas to display the models and animations on show. A network of guiding curves was attached to the pavilion's internal ceiling and floor in specific points, unfolding a series of minimal surfaces from stretch fabric to create projection screens, paravents and plafonds. The resulting continuous spatial network was translated into two material systems: a black structural space frame made from CNC milled guiding rails out of hard foam that were coated in Polyurethane as well as white suspended textile membranes. The two materials relate with their specific shape to their structural properties.

The exhibition thematically explores a series of research agendas conducted by Zaha Hadid Architects in recent years. Different media was used to show the work: architectural models, silver painting and projections.

A variety of projects from all over the world were shown, these included: the Soho Central Business District in Beijing, the Spiralling Tower of the University Campus in Barcelona, the Guggenheim project in Singapore, the recent completed CGM-CMA Tower in Marseille and the Pierres Vives building of the department de l'Herault in Montpellier, currently under construction. The exhibition also showcased architectural projects from the Arab world such as the Abu Dhabi Performing Arts Centre in the United Arab Emirates, the Nile Tower in Cairo Egypt, the Signature Towers in Dubai and the Rabat Tower in Morocco. Furthermore the exhibition showcases Zaha Hadid Architects' design research within the parametric paradigm.

Individual elements such as massing, skin, core, void, and structure were modulated individually and in concert. The final result was a fully malleable system that can differentiate families and fields of towers in response to user input or environment considerations. Applications of the research into architectural practice were exemplified via a series of Tower competition entries on large urban scales.

The visitor was invited to experience the work of Zaha Hadid Architects on three different levels, by discovering the Mobile Art Pavilion (building), viewing the exhibition design (scenography) and seeing the work of the practice (exhibits).

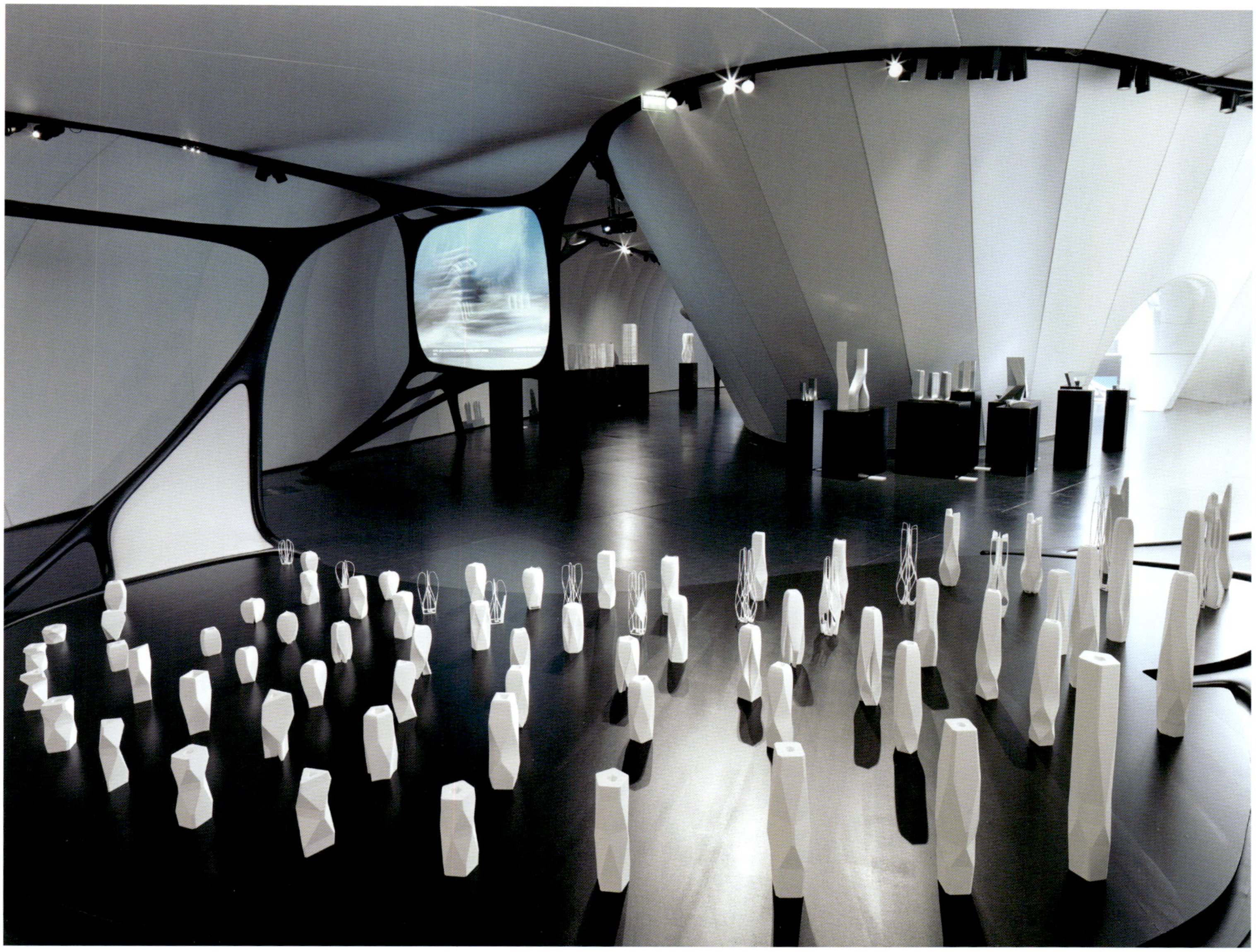

_SIGNITURE **TOWER**

_CMA-CGM **TOWER**

WORK*shop* _ISSUE.**TWO**

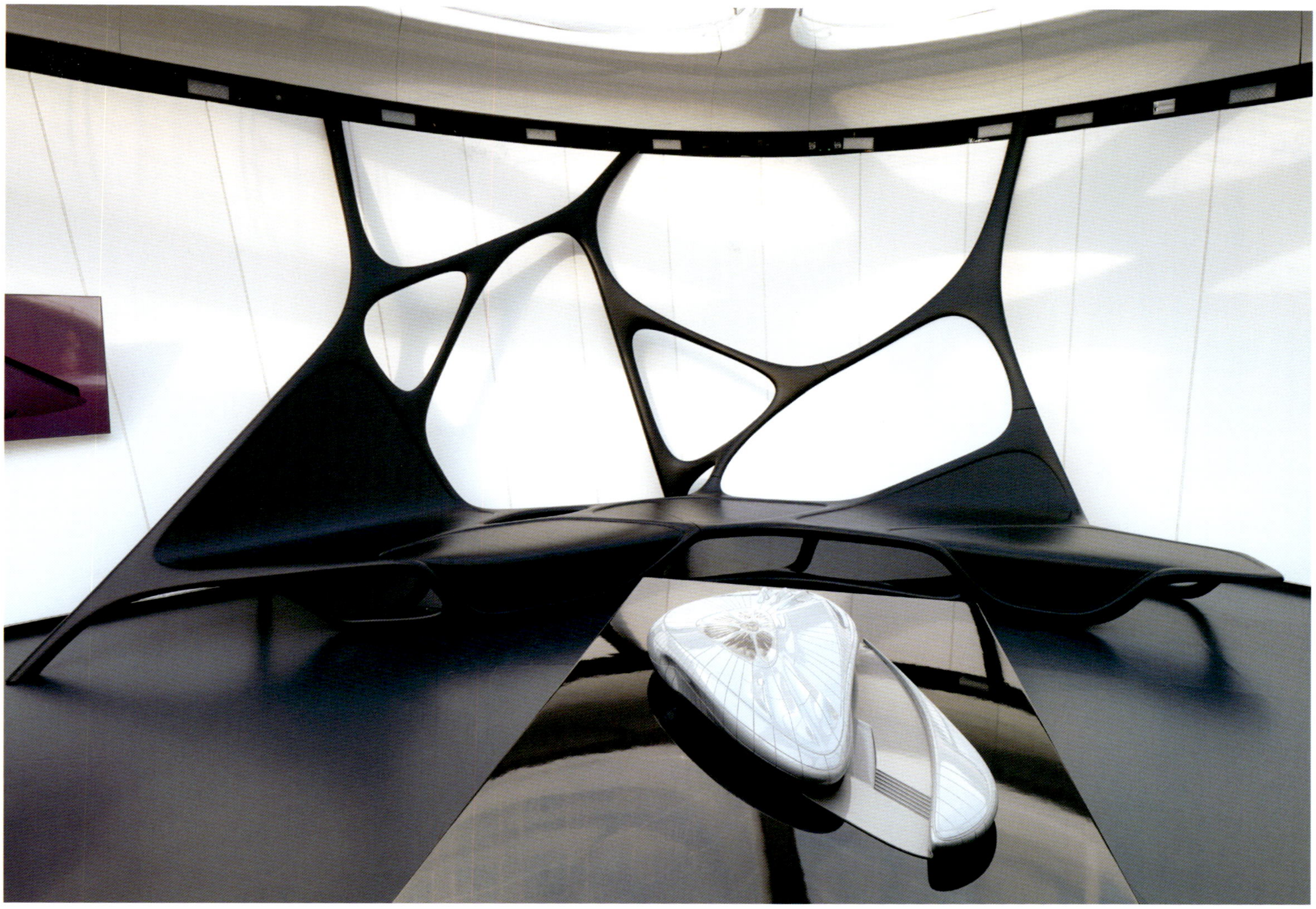

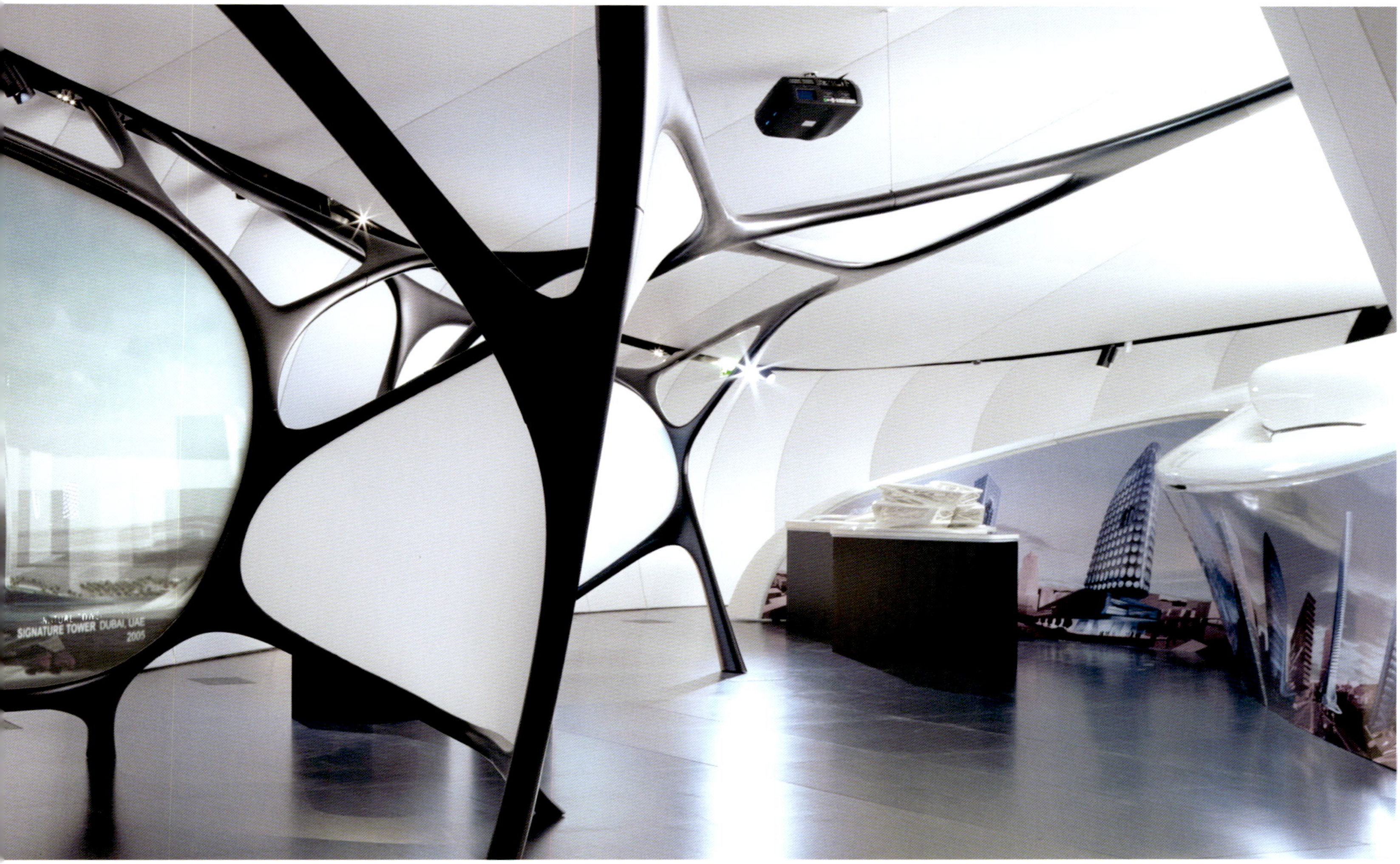
SIGNATURE TOWER DUBAI UAE
2005

ROCA LONDON GALLERY

_Zaha Hadid

The most recent Roca Gallery opened in London on 13 October 2011, designed by Zaha Hadid Architects, the Roca London Gallery is located in Chelsea Harbour and consists of a single floor measuring 1,100m². The leading role is played by the water that, in the words of Zaha Hadid, 'acts as a transformer moving, without interruption, through the facade, carving the interior and flowing through the main gallery as drops of water'. A flowing, all-white space made of faceted GRG (gypsum) panels and filled with drops of light serves as a central axis of the Gallery. Around this a number of smaller connected semi-enclosed spaces can be viewed through openings in walls. As a result, the visitor never feels enclosed in one space, but can always see beyond it into the space through overlapping and cutaway forms that enable a pleasing permeability to the Gallery.

The grey façade has a series of openings that seek to show the effect of water erosion. These drops of water connect the different areas of the space which are part of the Roca London Gallery itself; namely an exhibition of Roca's most innovative products, the meeting room and the multimedia space, amongst others. The continuing variety within the five distinct bathroom areas evolves along the backbone of the main gallery.

Here the visitor can find the most emblematic products displayed in a variety of areas and environments.

All the panels, which are made of GRC, or fibre reinforced concrete and extend up to 2.20m in height, have been pre-fabricated in moulds and constructed on-site. The façade is made of 2×4m panels of 800kg each. The panels creating the interior walls are 6cm thick and made of two waffled concrete layers sandwiching a honeycomb mesh that can stress in different directions and is very robust as a composite material. The furniture is made from GRP, or reinforced plastic, including the cove-shaped reception desk. The lighting scheme created by isometrics is also innovative in a complementary way, with special features including washing the walls in light and a mix of direct and dispersed mood lights.

A series of bathroom product ensembles are integrated in the space, and the Gallery's walls give way in six locations to semi-enclosed, cave-like spaces of GRC panels for the product displays, as well as to the bar and reception area. The cocoon-like meeting space has a wall of GRG, a continuation of the Gallery's central axis. A special feature of the Gallery is the floor of the product exhibition areas, which has a

EXHIBITION / SHOWROOM Design_ Zaha Hadid Photography_ Roca Country_ UK Client_ Roca

WORK*shop* _ISSUE.TWO

mosaic of porcelain tiles designed exclusively for the space by Zaha Hadid Architects. With each one cut and laid individually, the design creates an optical effect inspired by a water current.

New technologies and social media have become the primary communication tool today and for the visitor to the Roca London Gallery they play a key role. These new media resources enable visitors to be entertained as well as to learn about the values, history and special achievements of the brand, including its social responsibility and commitment to the environment and to water sustainability. Touchscreen interactive technology sited at the entrance is treated by Zaha Hadid Architects' design as an integral and complementary element within the Gallery.

The Roca London Gallery is intended to be much more than just a display space. Available to an extensive audience that will include everyone from design-savvy architects to design-hungry students, it will become a London hub hosting a wide range of activities such as exhibitions produced in-house or externally, meetings, presentations, seminars and debates, the criteria being a celebration of design in keeping with the Roca brand and company values.

ROCA LONDON GALLERY
GENERAL PLAN

PLAN

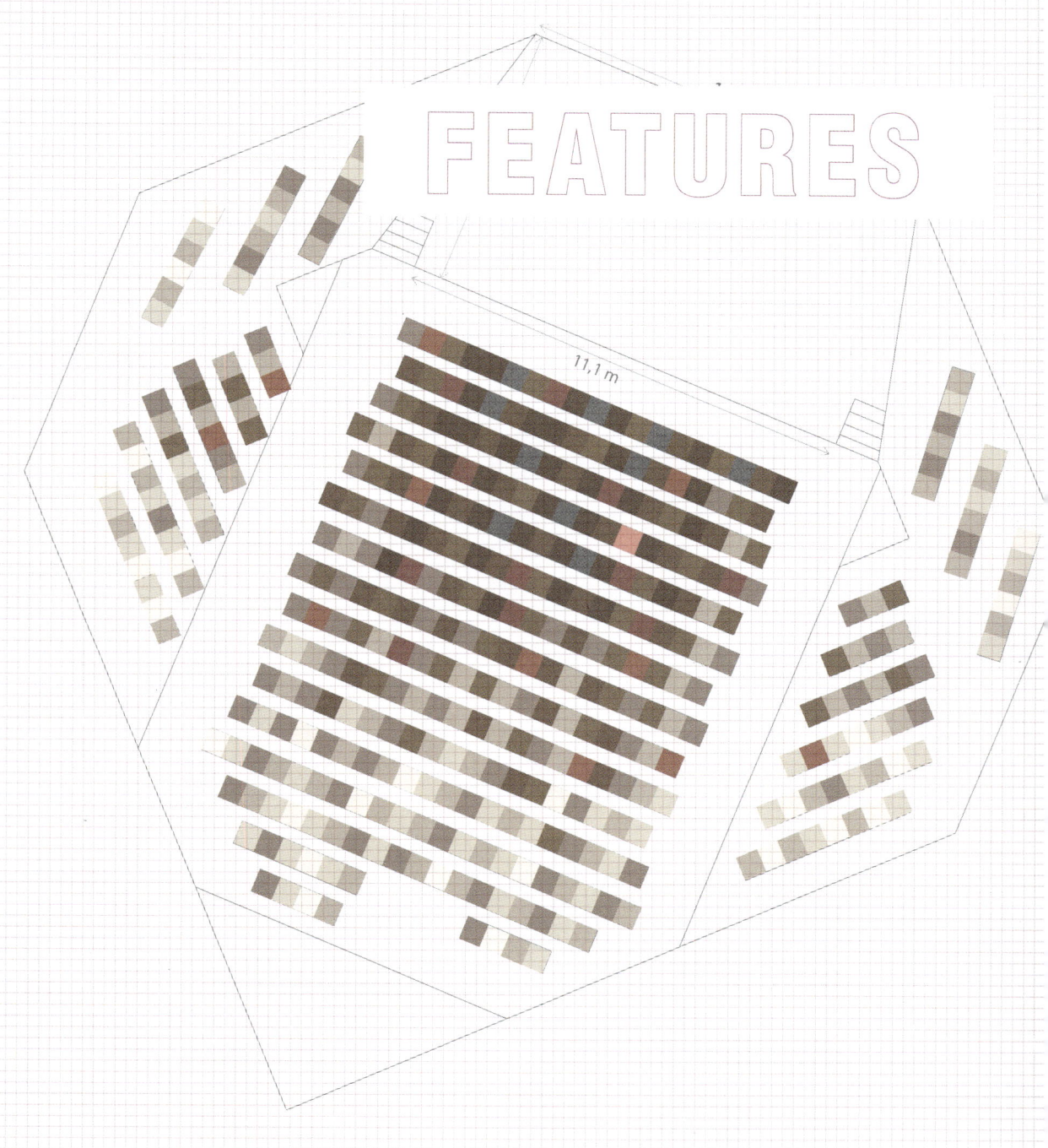

MUSIC HALL EINDHOVEN

DESIGN_VAN EIJK & VAN DER LUBBE TOGETHER WITH PHILIPS AMBIENT EXPERIENCE DESIGN (CREATIVE CONCEPT & DIRECTION)
INTERIOR ARCHITECT_VAN EIJK & VAN DER LUBBE
PHOTOGRAPHY_FRANK TIELEMANS

In the middle of Eindhoven now stands the absolute music centre of the future, a place where lighting, design and technology are integrated innovatively, without it becoming merely a high-tech building.
Niels van Eijk and Miriam van der Lubbe of Geldrop designed both the interior and the exterior around the central idea of the Concert Hall as a meeting place. Together with Philips Ambient Experience Design, Hypsos and Rapenburg Plaza, the duo produced an exceptional composition of light, image and specially developed technology. Take for example the way concert-goers are led intuitively from foyer to concert hall by way of subtle lighting signs which move from a high-tech wall and over the ceiling.
Van Eijk & Van der Lubbe designed every detail especially for this project, from the enormous glass façade to the foyers and furniture, from the working wear to the crockery.
The most remarkable change is to the main entrance of the new Concert Hall. This consists of a forward-leaning glass façade, 25m wide and 13m high. Behind it is a cultural city-foyer where people are welcome throughout the day for a cup of coffee, and to listen to and buy music. The city-foyer will be fitted with an ambient wall, several meters long, consisting of thousands of led lights, on which films, works of art and concerts will be projected. Visual artist Gerard Hadders realised the content of this living wall.
Visitors can listen to their favourite music in 'listening chairs' with integrated audio systems, designed specially by Van Eijk & Van der Lubbe. They also designed multi-functional furniture for the foyers, and special duo-chairs with an innovative lighting system.

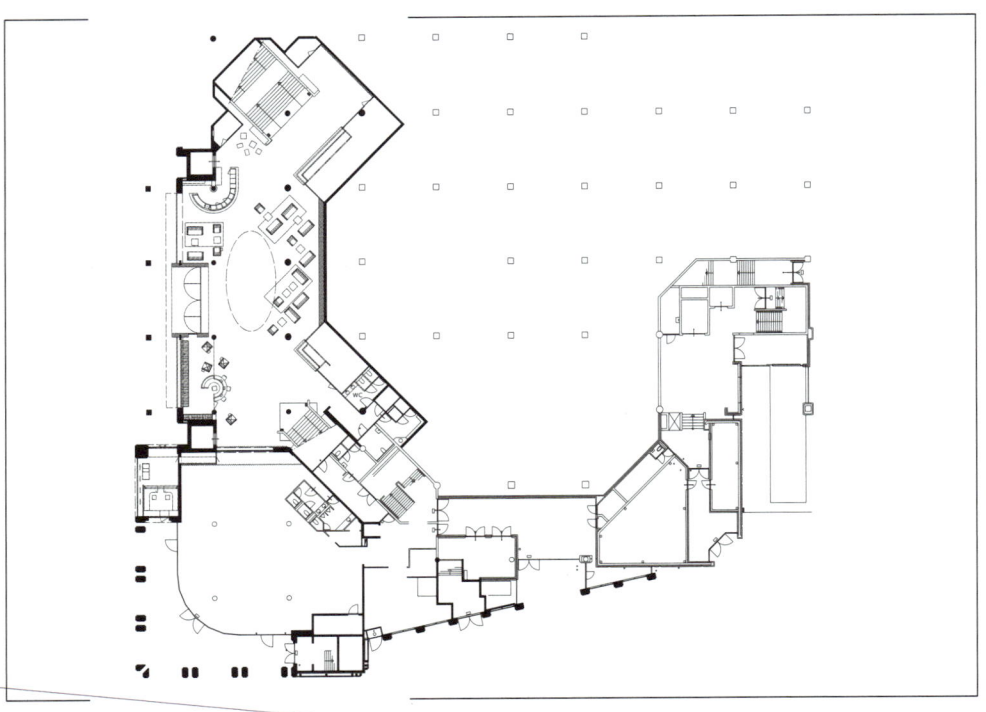

◀ GROUND FLOOR PLAN

▼ GROUND FLOOR

COUNTRY_THE NETHERLANDS

▲ FIRST FLOOR

SECOND FLOOR ▶

▲ DRESSING ROOM

*LIGHTING DESIGNER_*
RAPENBURG PLAZA

◀ CONCERT HALL

GROSS FLOOR AREA_
11,000M²

*AUDIOVISUAL SYSTEMS_*
MANSVELD GROEP PROJECTEN&
SERVICES

▲ CUP FOR ESPRESSO

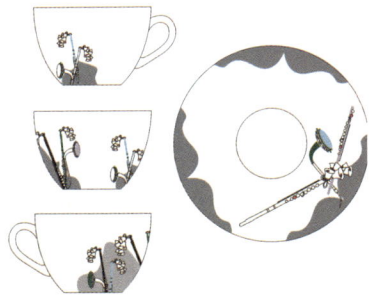

▲ CUP FOR COFFEE / CAPPUCCINO

It is a cultural city-foyer where people are welcome throughout the day for a cup of coffee, and to listen to and buy music.

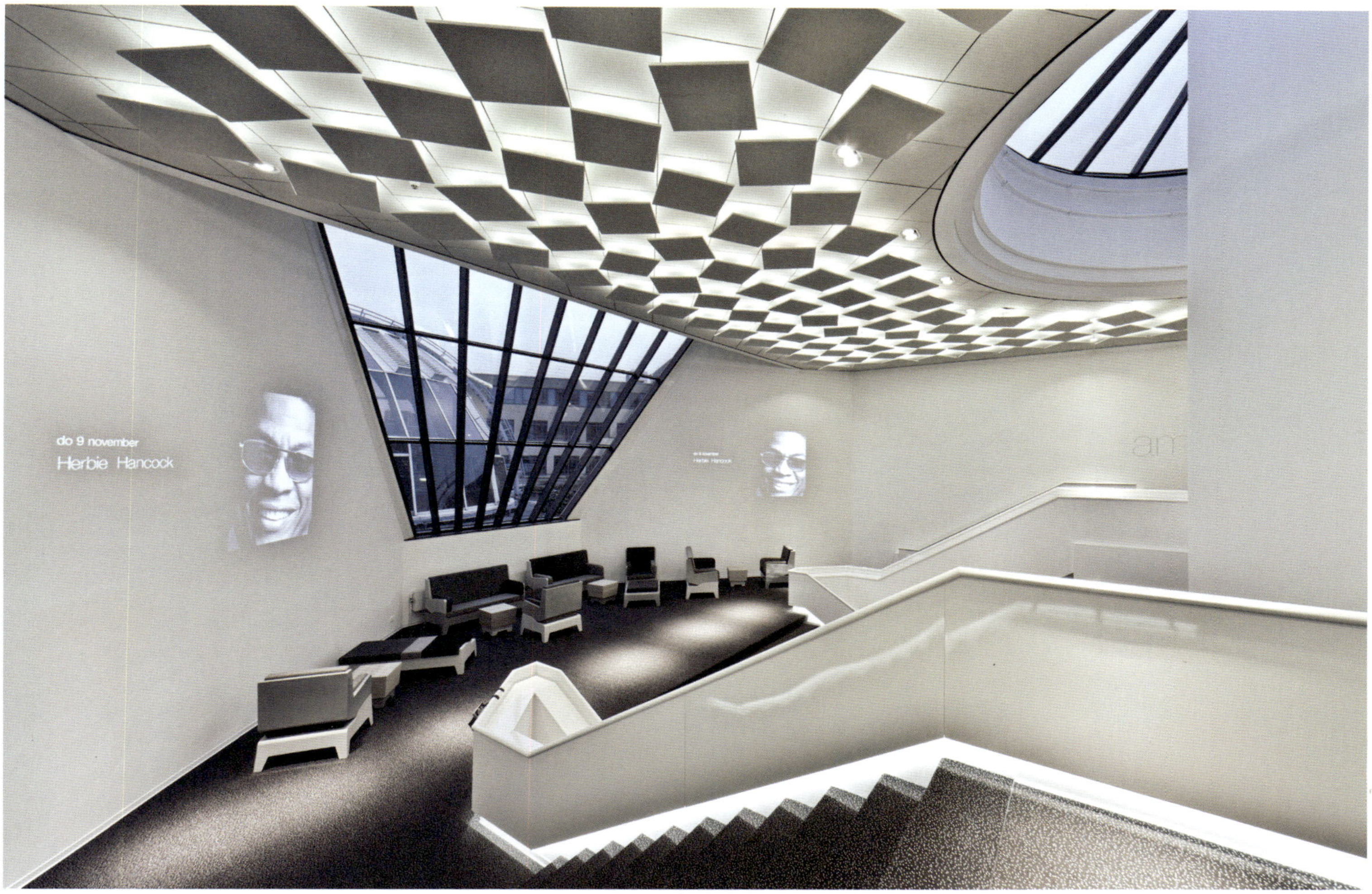
do 9 november
Herbie Hancock

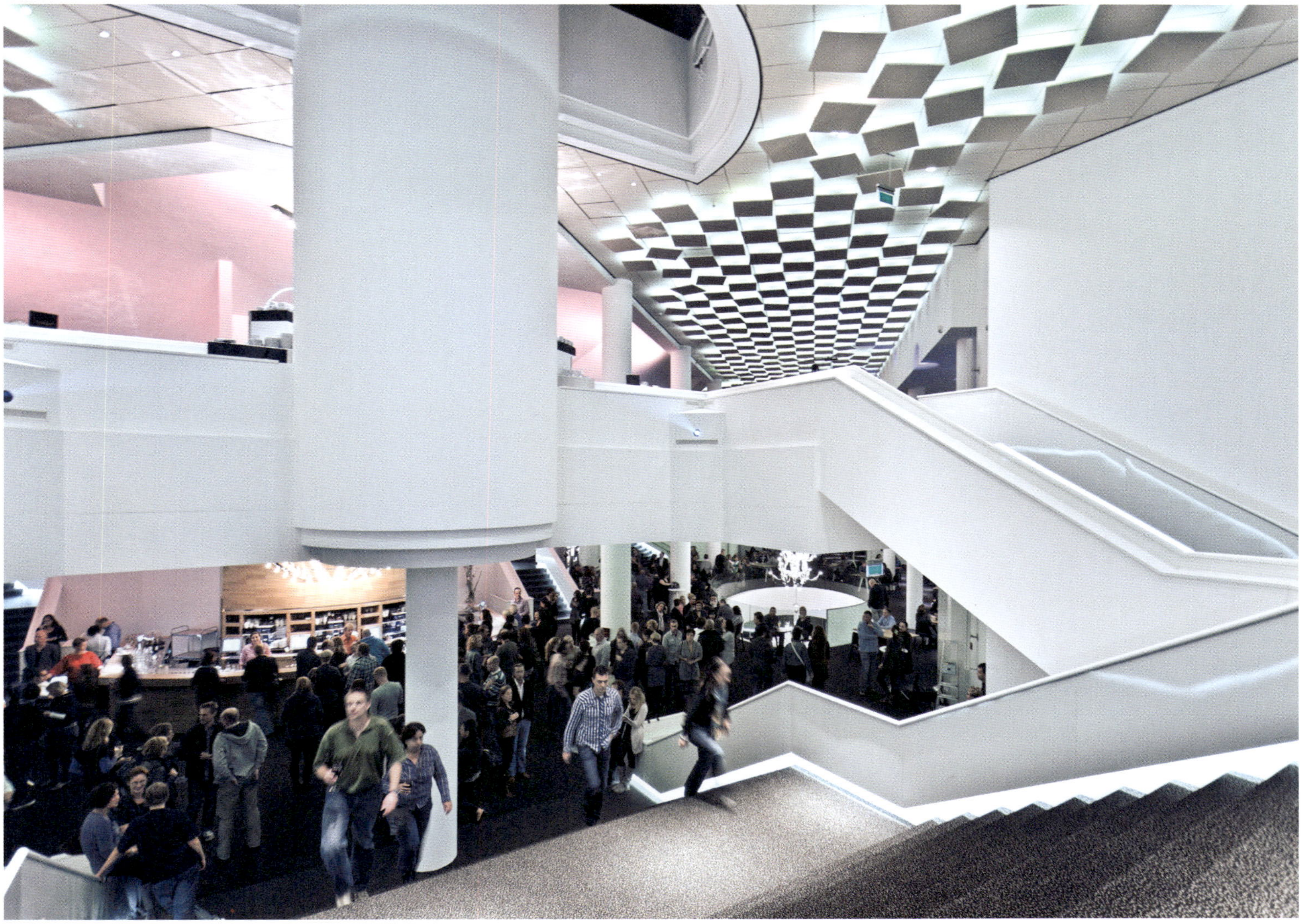

// Salon du Fromage

BIG CHEES O

RESTAURANT DESIGNED BY
KOTARO HORIUCHI ARCHITECTURE

// Kotaro HORIUCHI / Architect — Urbanist

COMPANY PROFILE

Kotaro HORIUCHI established his own firm, kotaro horiuchi architecture in Tokyo, Nagoya and Paris in 2009, after gaining experience through world renowned architecture firms such as: DPA – Dominique Perrault Architecture (France), Mecanoo Architecten (The Netherlands), and PPAG – Popelka Poduschka Architekten (Austria).

MISSION STATEMENT

As an up-coming young new generation architect. Kotaro HORIUCHI works on multiple projects around the world. His projects include both new construction and renovations of residentials, retails, offices, public housings, hotels, and spas.

TEAM

He manages offices in Tokyo, Nagoya, and Paris.

// Salon du Fromage

SALON DU FROMAGE

*LOCATION*_PARIS 1ST DISTRICT, FRANCE
*BUILT SURFACE*_73M²
*STRUCTURE*_BOLLINGER AND GROHMANN

A wall with curved surfaces was designed to softly wrap around the space on the ground and first floor of the existing building of the 18th century.
Piercing the space for selling the cheese on ground floor and the space for eating the cheese on the first floor in the existing building, the wall integrally changes to the space like the cave of the cheese.
In the very narrow space of the ground floor, refrigerated cylindrical display units with different diameters and heights, according to each 350 cheeses, are placed to let customers walk around to appreciate cheese in this space. Customers can anxiously select and taste cheese and they can find and savor themselves in the space.
Passing through a wall with curved surfaces and going up the stairs, one can find a restaurant on the first floor.
Open kitchen area is freely created to suit various applications by various cubical kitchen tables.
The space is furnished with tables and chairs and customers can select wines while enjoying a raclette or a cheese fondue.
The restaurant with an open kitchen has been designed to host various events such as cheese seminars, wine classes or cooking classes.

CHAISE PLIANTE

This chair was created for the 'Salon du Fromage'. Easy to be folded, just like a thin paper, this chair is made of four squares of 45cm side and 3mm thick. The chair is made of fiberglass to give strength to the structure.

MATERIAL_FIBERGLASS
COLOUR_WHITE, BLACK
DIMENSIONS_CHAIR / 45CM×45CM ×90CM

TABLE PLIANTE

MATERIAL_FIBERGLASS
COLOUR_WHITE, BLACK
DIMENSIONS_TABLE / 75CM×75CM ×75CM

This table was created for the 'Salon du Fromage'. Easy to be folded, just like a thin paper, this table is made of three squares of 75cm side and 3mm thick. The table is made of fiberglass to give strength to the structure. The countertop of the table can be pierced with holes of various sizes that can be just covered with glass to match the decor, hold a plant in its pot, a raclette grill or a cheese fondue set with its portable cooking stove. (Cable can be easily passed through the holes and connected to a plug on the floor.)

LUMIÈRE COUPÉE

MATERIAL_CORIAN® (COMPOSITE OF ACRYLIC POLYMER AND ALUMINA TRIHYDRATE)
COLOUR_WHITE, BLACK
DIMENSIONS_10CM×10CM×10CM
20CM×20CM×20CM
30CM×30CM×30CM
40CM×40CM×40CM

This lighting was created for the 'Salon du Fromage'. Shaped like a cube, this lighting has many round holes of various sizes that create interesting lights and shadows in the room. The accumulation of these lighting reminds the holes in the walls and gives a unique atmosphere to the space.
Cut in 6mm thick thermoformed Corian®, it has a milky colour that let the light glow trough the material. The design, without either juncture or sharp angles, is soft to the eye.

MORE FUN FOR YOUR LIFE AND WORK

PRODUCTS DESIGNED BY
KOAN DESIGN

'The lake is covered by a veil of mist and drizzle like a young maiden with thin white gauze, tears dripping down her chin and gently rippling the lake.'

Transform your desk into this landscape by the lake. Create a relaxing atmosphere full of calmness. Set your mind free and escape from daily stress. Put a branch of natural form into the rings like a tree growing in the middle of the lake – or simply use it as a penholder and discover how the pen drizzles into the lake.

KOAN CLOCK

When the clock runs without clock hands, the form of time becomes variable.
With this concept the designer changed the structure of the clock movement, and came out with a clock needing no clock hands.
Designer leaves the time a plate to draw its own pattern that changes its posture while time goes around.

PURPOSE_**DESK / WALL CLOCK**
SIZE_**Ø100*H137MM**
PACKAGE SIZE_**L150*W195*H120MM**
MATERIALS_**ABS**
PATENT PENDING_**098214880**
DESIGN_**KUO YAO HUANG**
DESIGN AND MADE IN TAIWAN

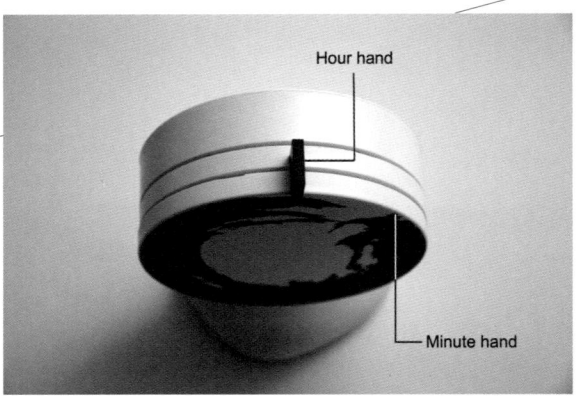

On the wall

10:20 02:50 03:00 10:10

KOAN + / KOAN DESIGN

Koan Design was established in 1999, as an interior design firm. The firm has extensive experience in residential, hotels, offices, and other commercial projects. In the recent years, it has won AsiaPacific Interior Design Awards and Taiwan Interior design award. That has given it confidence to extend its interest in product design. 'Koan+' is the new brand for its product, which was launched in summer 2009. The design concepts of the products are based on Asian origin culture and nature element.

BETWEEN

PURPOSE_**PENHOLDER**
SIZE_**D80*W200*H90MM**
PACKAGE SIZE_**L145*W260*H100MM**
MATERIALS_**ABS, ACRYLIC, STEEL**
DESIGN_**KUO YAO HUANG**
DESIGN AND MADE IN TAIWAN

RAIN TIMES

Rain comes oh rain!
So unexpectedly, it's such a pain
Grabbing a newspaper so you don't get your head wet
But hey now it's even better 'cos I have my Rain Time' bag
Rain comes oh rain!
Now my 'Rain Time' bag will always be in hand
Comes snow, sun or rain I'll be readied for any event!

PURPOSE_**TOTE BAG**
SIZE_**W470*H320MM**
PACKAGE_**OPP**
MATERIALS_**100% POLESTER**
DESIGN_**KUO YAO HUANG**
DESIGN AND MADE IN TAIWAN

T-FISH / V-FISH

PURPOSE_**DRINKING CUP**
SIZE_**Ø108*H130MM(T) / Ø80*H152MM (V)**
PACKAGE SIZE_**L150*W195*H120MM(T) / L150*W195*H120MM(V)**
MATERIALS_**PORCELAIN, PTR**
DESIGN_**JOYCE KUO**
DESIGN AND MADE IN TAIWAN

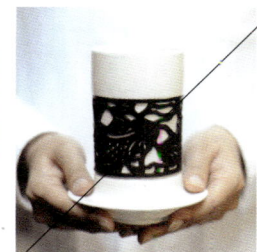

A gentle layer between you and the cup like a dance in front of your eyes, a whisper towards your fingertips. Hold the cup with both hands and feel the warmth in your palms.

T-BAMBOO / V-BAMBOO

PURPOSE_**DRINKING CUP**
SIZE_**Ø108*H130MM(T) / Ø80*H152MM(V)**
PACKAGE SIZE_**L150*W195*H120 MM(T) / L150*W195*H120MM(V)**
MATERIALS_**PORCELAIN, BAMBOO**
DESIGN_**PATRICK CHEN**
DESIGN AND MADE IN TAIWAN

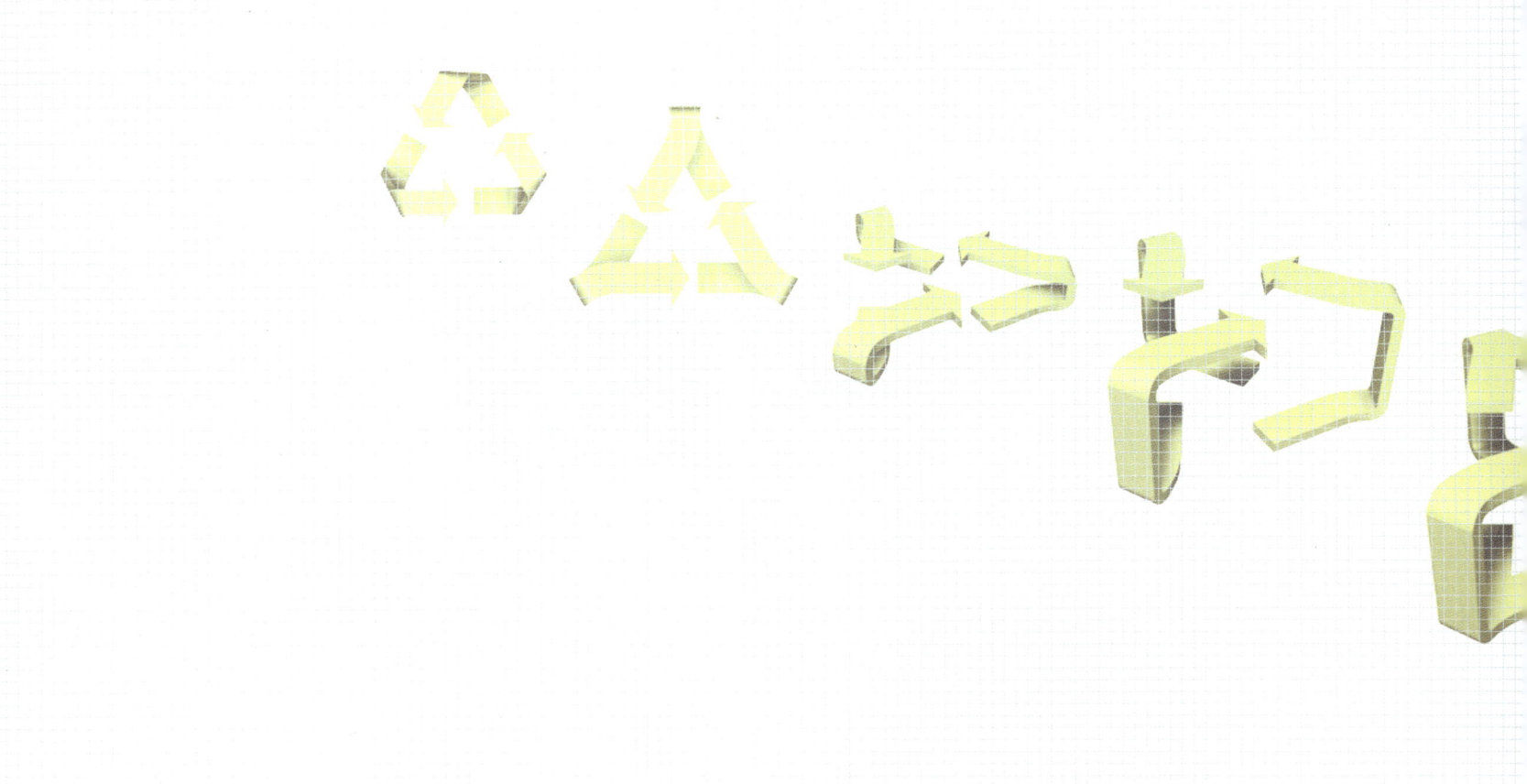

DESIGNER

STREAMLINE ICON

Based in Berlin, J. Mayer H. Architects is a multi-disciplinary design office with portfolio work in the fields of architecture, interiors, installations, product design, and furniture. J. Mayer H. Architects' works re-elaborate the architecture by their work and German enthusiasm, not just as people usually think about German design – thick and grey. Through balance between envelope-pushing forms and wild spatial experimentation with thoughtful and appropriate concepts, the dynamic and streamline projects demonstrate here, while visually striking, performs just as well when considered in a programmatic or technical sense.

interview

Interview with Jürgen Mayer H.

Interviewer: Choi's Workshop (WORKSHOP)
Interviewee: Jürgen Mayer H., founder and principal of J.Mayer.H. Architects (J.M.H.)

Photography_Jens Passoth

WORKSHOP: Please briefly introduce yourself and your office.
J.M.H.: Founded in 1996 in Berlin, Germany, our studio focuses on works at the intersection of architecture, communication and new technology. Recent projects include, a student centre at Karlsruhe University, the villa Dupli.Casa near Ludwigsburg, Germany and the redevelopment of the Plaza de la Encarnacion in Sevilla, Spain, the office buildings ADA1 and S11 in Hamburg, Germany and the extension of the science park in Danfoss, Denmark. From urban planning schemes and buildings to installation work and objects with new materials, the relationship between the human body, technology and nature form the background for a new production of space. I founded this cross-disciplinary studio. I studied at Stuttgart University, the Cooper Union and Princeton University. My work has been published and exhibited worldwide and is part of numerous collections including MoMA New York and SF MoMA. National and international awards include the Mies – van – der – Rohe – Award – Emerging – Architect – Special – Mention – 2003, Winner Holcim Award Bronze 2005 and Winner Audi Urban Future Award 2010. I have taught at Princeton University, University of the Arts Berlin, Harvard University, Kunsthochschule Berlin, the Architectural Association in London, the University of Toronto, Canada and I am currently teaching at the Columbia University, New York.

WORKSHOP: You do pay attention to relationship between the human body, technology and nature, is that the motif of your design? Is there any other motif?
J.M.H.: One major investment in our work is looking at expanding the material of architecture, beyond just building material. The influence of new media and new materials now expands our understanding of 'space' as a platform for communication and socio-cultural interactivity.

WORKSHOP: What do you think is the biggest challenge your design will meet in the future? In 10 years maybe.

J.M.H.: I am very curious to see how our speculations about communications and public space might transform in future once they are handed over and begin their own life. The town-hall remains the wonder-box with most ingredients already prepared. From thereon, our current projects develop specific aspects in different directions, testing the possibilities of architecture.

WORKSHOP: When it comes to future, what do you look forward to the young generation, since we know that you've taught at several universities?
J.M.H.: Teaching is an extremely important element in my discourse on architecture. I am mainly interested in how cultural phenomena condense on architecture, frame new challenges in how we produce and look at architecture, and how we can speculate about the future role of architecture. Columbia University, where I am teaching right now, has recently developed into a breeding ground of intense enquiries toward that uncertainty of what architecture is. Students become scouts and specialists, or even better 'speculationists'. What can be tested in a semester as a theoretical thesis, is what in parallel concerns us in our practice.

WORKSHOP: Let's come back to design, in the aspect of material application, do you have any preference? And why?
J.M.H.: Part of the research we do is with companies at the forefront of material development. New programmes, new requirements like sustainability, atmospheric demands and even duration and lifetimes, ask for new construction methods and a more complex performance of materials. For me it is interesting to see what happens if you use new materials and traditional ones to challenge conventional understanding of space. Our temperature-sensitive furnitures like Heat.Seat show very clearly that there is no innocent surface anymore. We leave traces wherever we are and this information has a certain value.

WORKSHOP: How do you define 'Design' and manage it in a specific project?

J.M.H.: Design is an ongoing process. We try to establish parameters as a skeleton or framework for each project. These are conceptual conditions rather than design driven compositions, based on a client's brief, contextual references and programmatic logistics.
For example: we established a 3D diagram called the gum.gram or nutella.gram. Imagine two pieces of bread with a thick paste between them. We accepted the given site layout as the footprint of the building and only looked at the relationship between the floor and the roof. By compressing the functional programme between those two layers quite naturally the position of the roof ended up on a strange angle, thicker and closed volume at the rear side, thinner and more open to the front. By keeping to the reference to an elastic space we were not fixed to a specific volume. The structural elements became references to the diagrammatic process as well as to the blurry nature-urban transition that defines the site. Now, after working on the material aspects, we translated this elasticity into a timber structure coated with a specially developed polyurethane skin which has a strange rubbery quality. This structural solution has become a prototype for public buildings because of its sustainability qualities, low maintenance and costs. Right now it is the largest timber construction site in Germany. The elasticity concept translates all the way to the materialization of the building.

WORKSHOP: How do you tackle with the relationship with your clients? Have you ever encountered a serious situation that you're forced to change your original design?

J.M.H.: If you develop certain questions about the potential of architecture and how it generates a dialogue between a client's brief, context and an inherent architectural discussion, you will find that scale is increasingly unimportant, in other words, to investigate certain concerns in different scales cross references from the object to the urban context.

WORKSHOP: Please talk something about your main and most important project, which may underline your vision today.

J.M.H.: Metropol Parasol is our most important project since years. Based on an archaeological window into the history of Sevilla, the parasols cover the very heart of the city for a new urban place for the 21st century. Metropol Parasol covers history, hosts the everyday life in the food market or in commercial spaces, offers open public space for events and contemplation, and it refers to a 'visionary' culture with rising structures to elevate visitors onto a panoramic Sevilla city view on the roof-scape. All these different programmes are open and active at various times of day and night. Actually Sevilla as with most Spainish cities, is very close to a 24 hour urban space. Whenever you walk around in the city, there is a lively, energetic atmosphere. As much as the parasols provide shadow during the day, night-time might become even more important when Metropol Parasol creates an atmospheric cover to various forms of public activities still to be invented.

WORKSHOP: What's the biggest difficulty you've ever met, and how did you get through it?

J.M.H.: After almost three years of working on a Spanish project I still wonder why some things are extremely fast and euphoric, and some other things are so bureaucratic and slow. I see a lot of curiosity and pride, which I believe drives all levels of cultural production, from theatre to dance, fashion and art. It is a moment of exploration. Architecture in Spain is considered a great cultural adventure – actually not surprising if you look at their rich inventive architectural history.

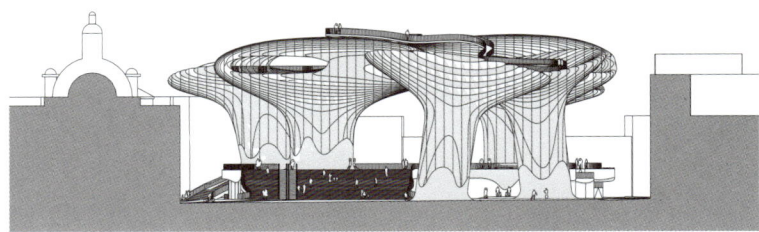

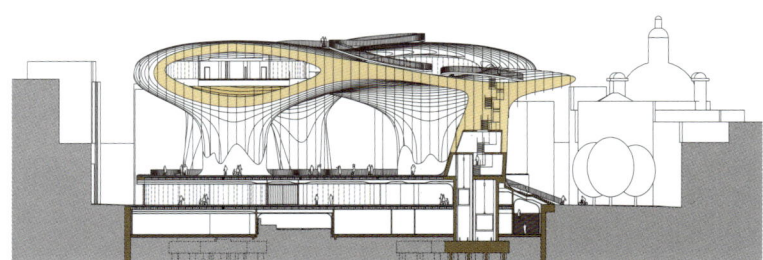

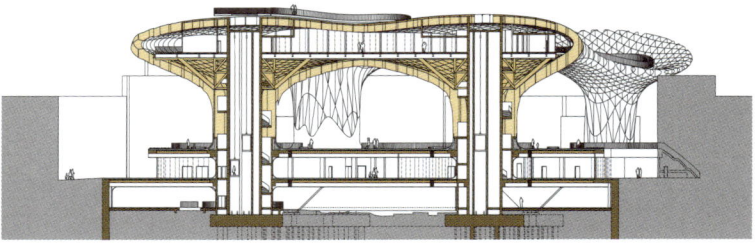

SECTIONS

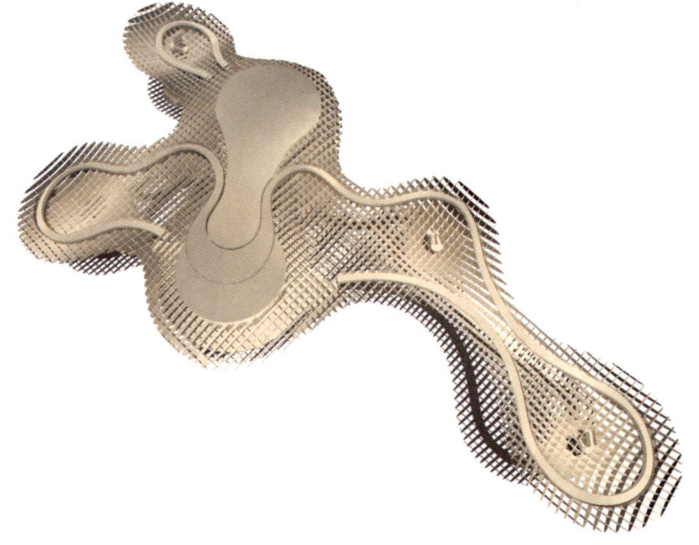

ROOF

METROPOL PARASOL

DESIGN_	J. MAYER H. ARCHITECTS_JÜRGEN MAYER H., ANDRE SANTER, MARTA RAMÍREZ IGLESIAS
DESIGN TEAM_	ANA ALONSO DE LA VARGA, JAN-CHRISTOPH STOCKEBRAND, MARCUS BLUM, PAUL ANGELIER, HANS SCHNEIDER, THORSTEN BLATTER, WILKO HOFFMANN, CLAUDIA MARCINOWSKI, SEBASTIAN FINCKH, ALESSANDRA RAPONI, OLIVIER JACQUES, NAI HUEI WANG, DIRK BLOMEYER
LOCATION_	SEVILLE, SPAIN
CLIENT_	AYUNTAMIENTO DE SEVILLA AND SACYR
PHOTOGRAPHY_	FERNANDO ALDA
CONSTRUCTION AREA_	SITE 18,000M², BUILDING 5,000M², TOTAL FLOOR 12,670M²
HEIGHT_	28.50M
MATERIALS_	CONCRETE, TIMBER, GRANIT AND STEEL
COST_	90,000,000
PHASE_	COMPETITION 2004	PROJECT 2004-2011

METROPOL PARASOL

Spring 2011 marks the opening of 'Metropol Parasol', the Redevelopment of Plaza de la Encarnacon in Seville, designed by J. MAYER H. Architects. After finishing the concrete works in 2008, the parasols are under construction now. Visiting the site at the moment gives an impressive imagination of the final dimension and appearance.

The project becomes the new icon for Sevilla, a place of identification and to articulate Sevilla's role as one of Spain's most fascinating cultural destinations. 'Metropol Parasol' explores the potential of the Plaza de la Encarnacion to become the new contemporary urban centre. Its role as a unique urban space within the dense fabric of the medieval inner city of Sevilla allows for a great variety of activities such as memory, leisure and commerce. A highly developed infrastructure helps to activate the square, making it an attractive destination for tourists and locals alike.

The 'Metropol Parasol' scheme with its large structures offers an archeological site, a farmers market, an elevated plaza, multiple bars and restaurants underneath and inside the parasols, as well as a panorama terrace on the very top of the parasols. Realized as an innovative timber-construction with a polyurethan coating, the parasols grow out of the archeological excavation site into a contemporary landmark. The columns become prominent points of access to the museum below as well as to the plaza and panorama deck above, defining a unique relationship between the historical and the contemporary city. 'Metropol Parasol's mix-used character initiates a dynamic development for culture and commerce in the heart of Sevilla.

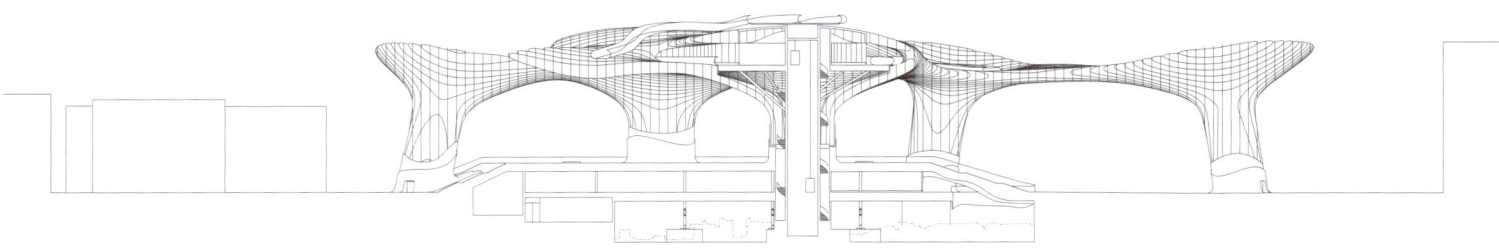

SECTION

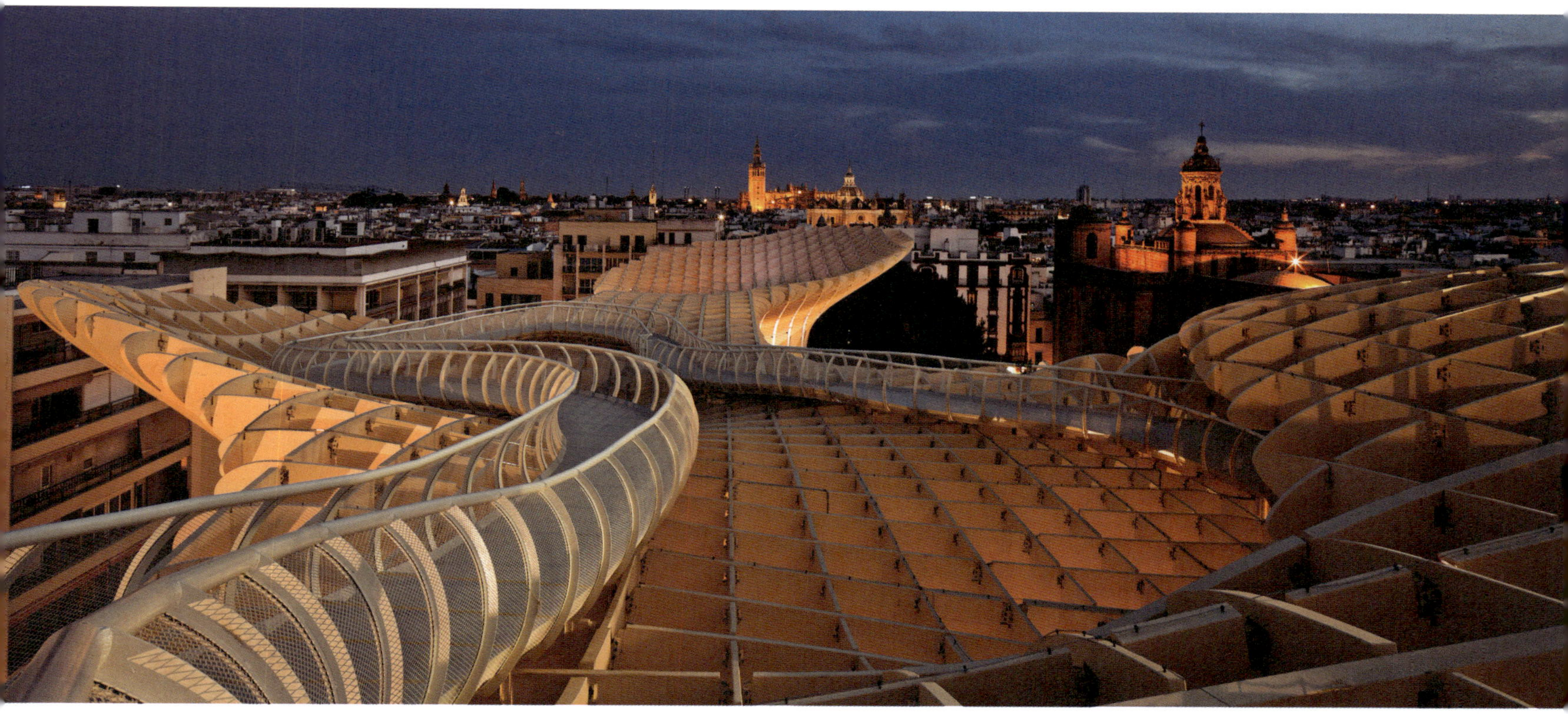

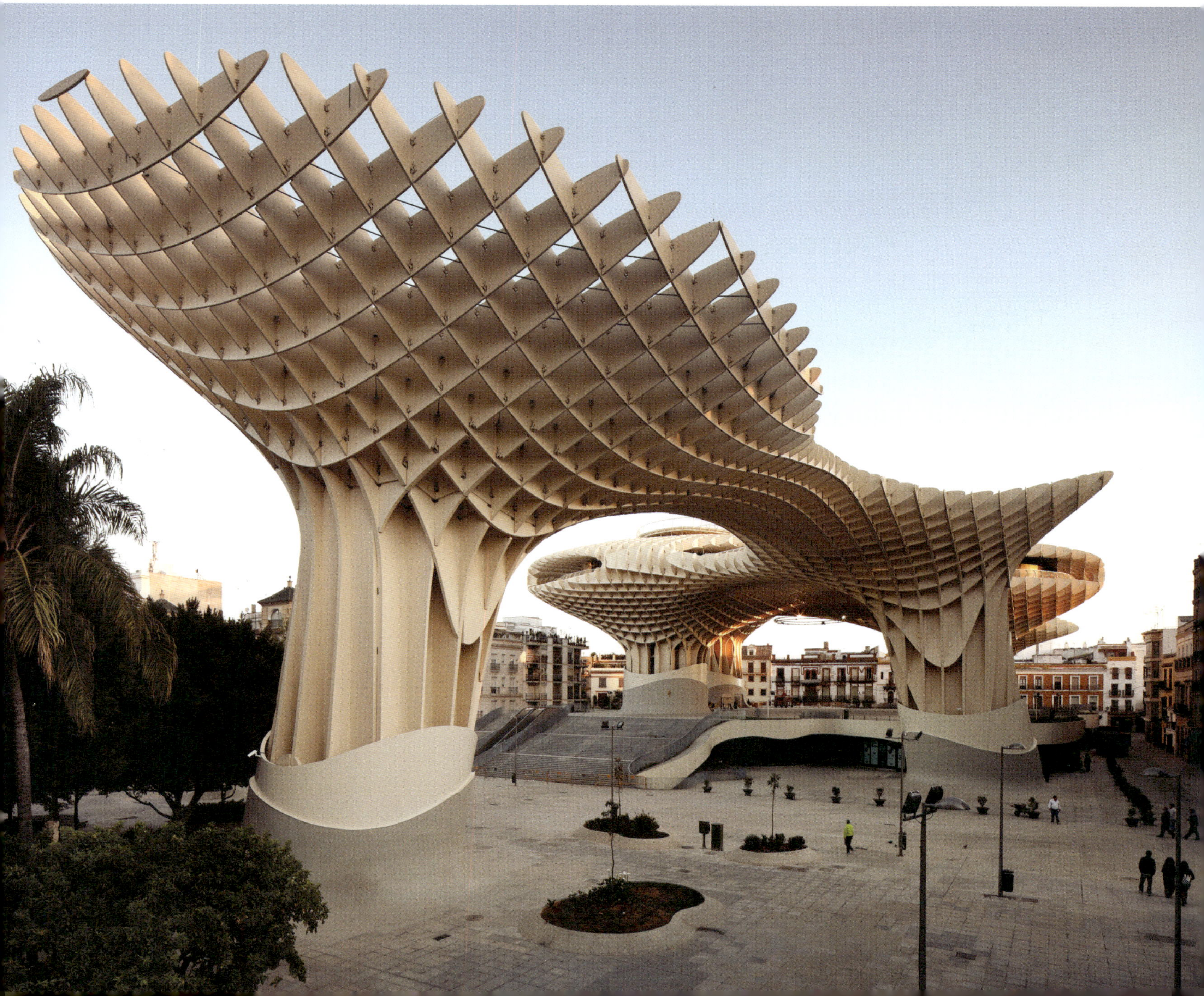

DANFOSS UNIVERSE

DESIGN_	J. MAYER H. ARCHITECTS
DESIGN TEAM_	JÜRGEN MAYER H., MARCUS BLUM, THORSTEN BLATTER, ANDRE SANTER, ALESSANDRA RAPONI
LOCATION_	NORDBORG, DENMARK
CLIENT_	DANFOSS UNIVERSE
PHOTOGRAPHY_	J. MAYER H. ARCHITECTS
EXHIBITION_	'BITLAND', SCIENCE EXHIBITION FOR KIDS
BUILDING AREA_	CAFETERIA 500M², CURIOSITY CENTER 1,200M²
SIZE_	CAFETERIA 46/10/11M, CURIOSITY CENTER 72/17.5/16M (LENGTH/HEIGHTS/WIDTH)
MATERIALS_	STEEL CONSTRUCITON COVERED WITH CORRUGATED SHEETMATEL
COST_	3,000,000
PHASE_	DESIGN 2004	CONSTRUCTION 2004-2011

DANFOSS UNIVERSE

The new buildings rise up from the ground and provide spaces which articulate the fusion of outdoor landscape and indoor exhibition. This active ground modulates according to program and location in the park. The endpoints of the buildings blur the line between building and park by offering inside-out spaces as display areas and projection surfaces related to the temporary exhibitions inside. Silhouettes, as groups of land formations, define the unique newly programmed horizon line of Danfoss Universe.

Danfoss Universe is a science park in Denmark, embedded in the agricultural landscape of Nordborg next to the founder's home and the Danfoss headquarter. It opened in May 2005 and is already enlarging due to its considerable success. The masterplan for Danfoss Universe Phase 2 includes an exhibition building (Curiosity Center) and a restaurant (Food Factory) which extends the summer based outdoor park into the winter months by enclosing spaces for exhibitions and scientific experiments.

SITE PLAN

CURIOSITY CENTRE SECTION

CAFETERIA AND MAIN ENTRANCE

CURIOSITY CENTRE

AN DER ALSTER 1

DESIGN_	J. MAYER H. ARCHITECTS
DESIGN TEAM_	JÜRGEN MAYER H., HANS SCHNEIDER, WILKO HOFFMANN, ANDRE SANTER, SEBASTIAN FINCKH, MARTA RAMÍREZ IGLESIAS, GEORG SCHMIDTHALS, MARCUS BLUM
LOCATION_	HAMBURG, GERMANY
CLIENT_	COGITON PROJEKT ALSTER GMBH, HAMBURG
PHOTOGRAPHY_	FOTOGRAFIESCHAULIN, SCHRAUBVERSCHLUSS
CONSTRUCTION AREA_	BUILDING 970M², (BGF)FLOOR 5,436.16M², NUTZFLÄCHE 4,512.03M²
HEIGHT_	20.50M
PRINCIPAL EXTERIOR MATERIALS_	AERATED CONCRETE AND GLASS, DOUBLE SKIN FAÇADE
PRINCIPAL INTERIOR MATERIALS_	CONCRETE AND PLASTERBOARD, WATER BASED COOLING SYSTEM INSIDE THE CONCRETE, NO AIR-CONDITIONING NEEDED
PHASE_	COMPETITION 2005	PROJECT 2005-2007

OFFICE BUILDING 'AN DER ALSTER 1'

The building site is situated at the intersection between Hamburg's lively downtown and its urban landscape that is rich in water and mature trees. It is at the transition from city to nature, and the gateway building to the bustling metropolitan core.
The horizontal striped façade with its floating 'eyes' celebrates the view onto this unique context. A public park in front of the building continues the design strategy of the façade into the landscape. The 'eyes' in the façade and the platforms in the park enable the places to meet and contemplate.
The office spaces serve both a generic spatial layout and specific moments related to the 'eyes'. Large spans provide for various office layout configurations in combination with balconies and climatically tempered outdoor spaces of the 'eyes'.
The 'eyes' are just a bigger space inside the double skin façade, which work the same as the whole façade.

This kind of façade highly economizes on energy, because the windows of the inner façade can be used for ventilation. This is the reason why no additional air conditioning is needed for the office building. In Germany, the companies mostly don't want buildings with air conditioning because of the high running costs. Furthermore an air conditioning is also problematic in terms of health conditions. In addition to that the concrete core activation system inside the concrete ceilings cools the building during summertime and heats it during wintertime with the help of water. The warmed surfaces allow lower heating costs in wintertime and thus save energy.
The office building 'An der Alster 1' links its interior and exterior spaces to the public park in front of the building and to the city context of Hamburg, becoming a new anchor at the prestigious Aussenalster waterfront.

MAIN ENTRANCE

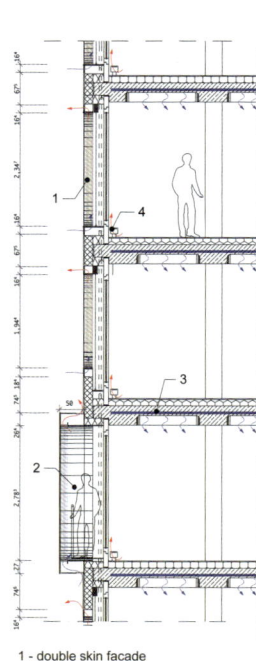

1 - double skin facade
2 - "eye" loggia inside the double skin facade
3 - concrete core activation system

SECTION

MAIN FAÇADE

BACKSIDE

SIDE VIEW

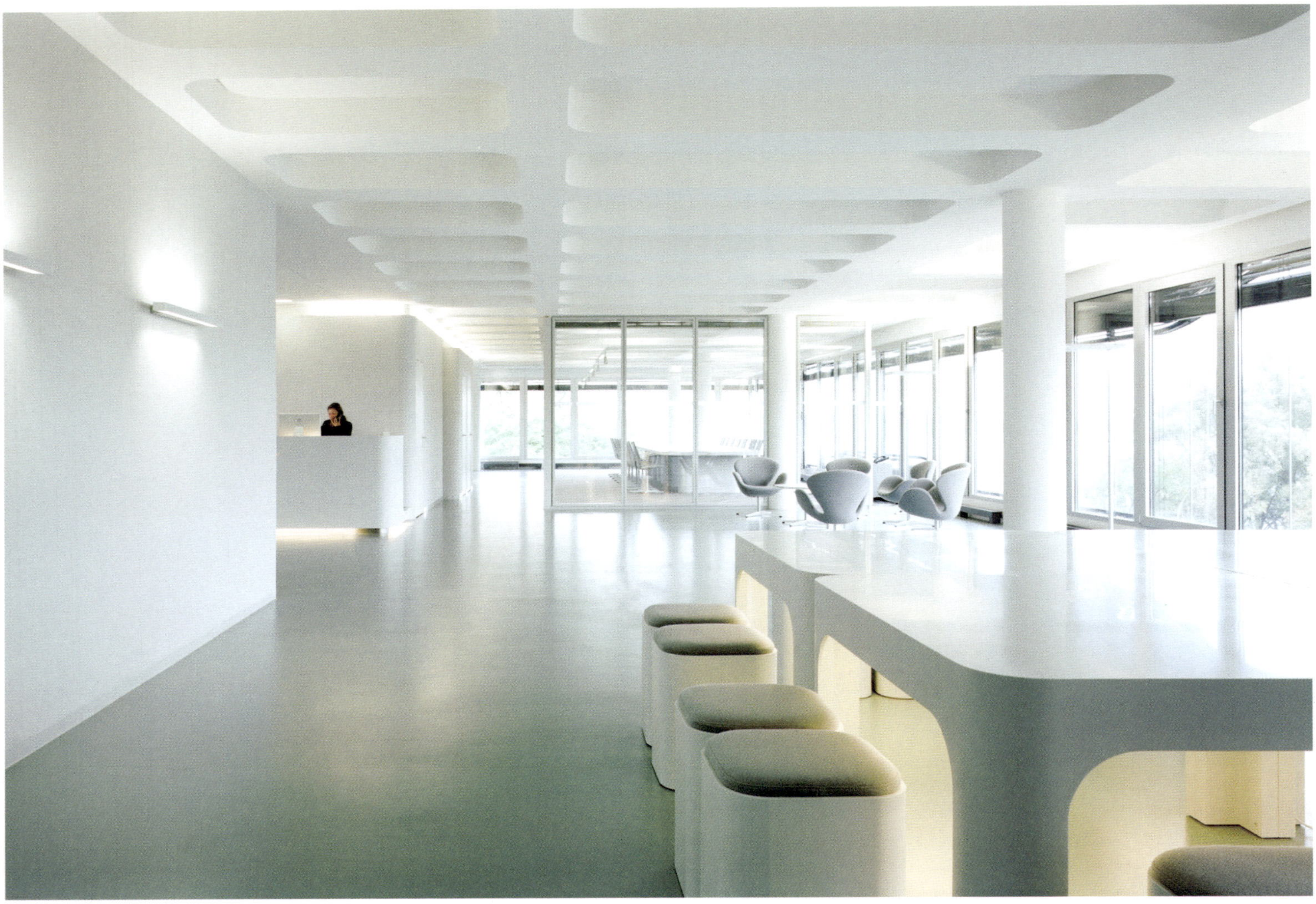

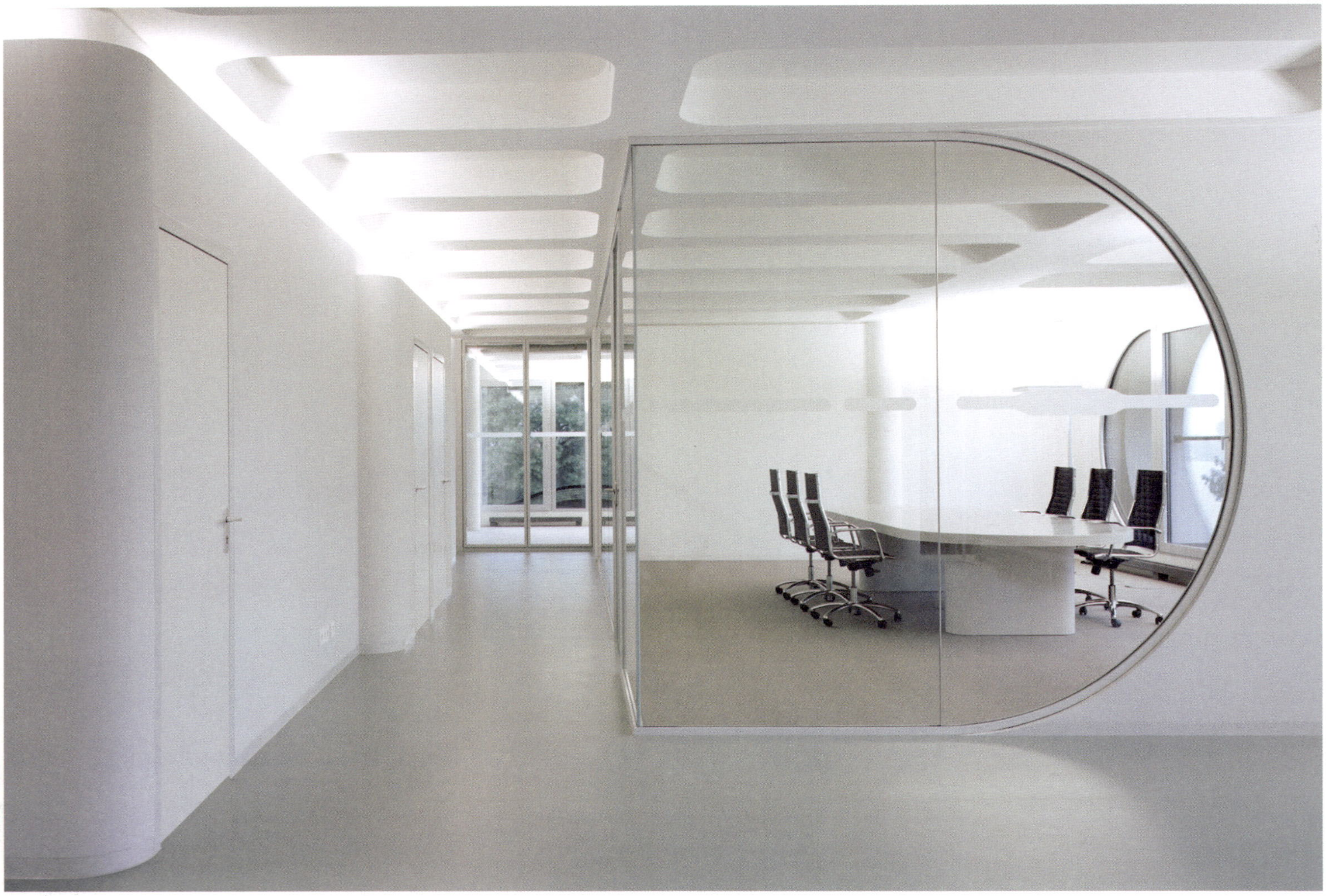

CORNER WITH 'EYE'

LEVEL GREEN

DESIGN_ J. MAYER H. ARCHITECTS
DESIGN TEAM_ JÜRGEN MAYER H., JAN-CHRISTOPH STOCKEBRAND, PAUL ANGELIER, MEHRDAD MASHAIE, JONATHAN BUSSE, DANIEL MOCK, STEFAN HENTRICH
CLIENT_ AUTOSTADT GMBH, WOLFSBURG
PHOTOGRAPHY_ AUTOSTADT GMBH
AWARD_ REDDOT DESIGN WWARD 'BEST OF THE BEST 2009', ANNUAL EXHIBITION SPACE AWARD, CHINA, 'WINNER'
TOTAL FLOOR AREA_ APPROX. 1,000M²
MATERIALS_ WOOD COMPOSITE WITH STEEL REINFORCEMENT
PHASE_ COMPETITION 2007 | PROJECT 2007-2009

EXHIBITION LEVEL GREEN
THE CONCEPT OF SUSTAINABILITY

The offices of J. MAYER H. Architects and Art+Com Berlin were commissioned to develop a permanent exhibition on the topic sustainability for the Autostadt in Wolfsburg, Germany. The exhibition Level Green was opened on the 4th of June 2009 and encompasses approximately 1,000m².

The architectural design of Level Green takes the numerous interdependencies of the topic as a starting point and translates this quality into the metaphor of the web. Similar to a continuous organism, the single elements of the exhibition are connected into one homogenous structure that houses all content and technical installations.

As one of the first prominent signs of the growing consciousness for environmentally friendly consumption, the well-known PET-sign was taken as a starting point from which the metaphor of the extensively branched web was developed. This originally two-dimensional sign was extended into the third dimension and through a series of step by step manipulations a complex structure was created, which allows for an abstract property of the topic to be experienced on a spatial level.

The dramaturgy of the exhibition is not determined by a linear approach but one of nonlinear logics, opening the space for a more ambiguous experience. Vertical Elements define different areas within the exhibition without strictly separating them, allowing the visitor's experience to be carried by the idea of playful discovery.

After a phase of extensive material research, the design was executed by the use of easily processed wood composite sheets (MDF) with varying thickness according to the structural and geometrical demands. The MDF-material is specially treated to meet the fire rating requirements (B1) necessary for this project. In order to guarantee the structural performance of the construction, all vertical elements were reinforced with a steel structure and bolted to the concrete floor. After various testing, the colour coating was executed with acrylic-based car paint, developed to guarantee high usability while meeting strict environmental regulations. The painting was done by local firms accustomed to high quality standards common in the automobile industry.

The concept for the display of information is based two main formats: the object-like display of data and statistics on a more sensational level and touch sensitive surfaces for in-depth explanations on different aspects of the topic. Designed to evoke the visitors initial interest, the first are placed within the exhibition space in the form of data sculptures or sample objects. The latter take on the form of black surfaces for interaction or information carrier and are seamlessly integrated into the vertical elements which define different areas within the space.

Necessary technical installations are also integrated into the design and appear only in the abstract form such as glowing lines or painted covers.

As far as the subject matter is concerned, the exhibition Level Green argues for scientific research and the use of latest technological development as necessities for survival in the future. This point of view is represented as an atmospheric environment, in which physical and digital spaces complement each other, creating one common narrative.

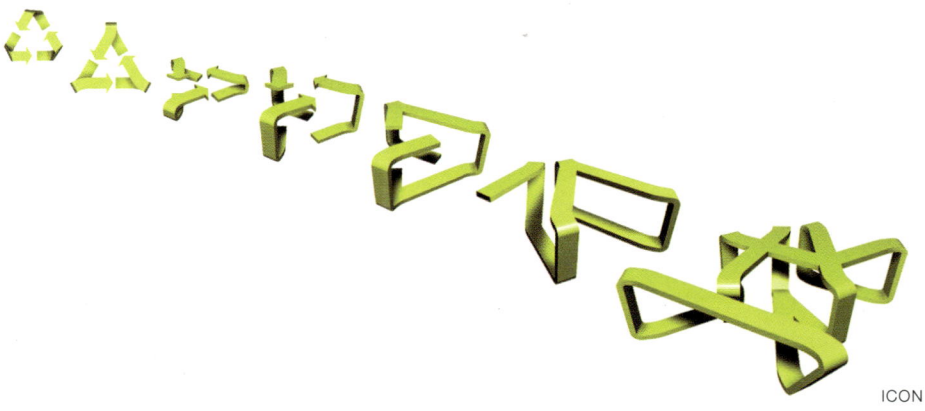

ICON

_247

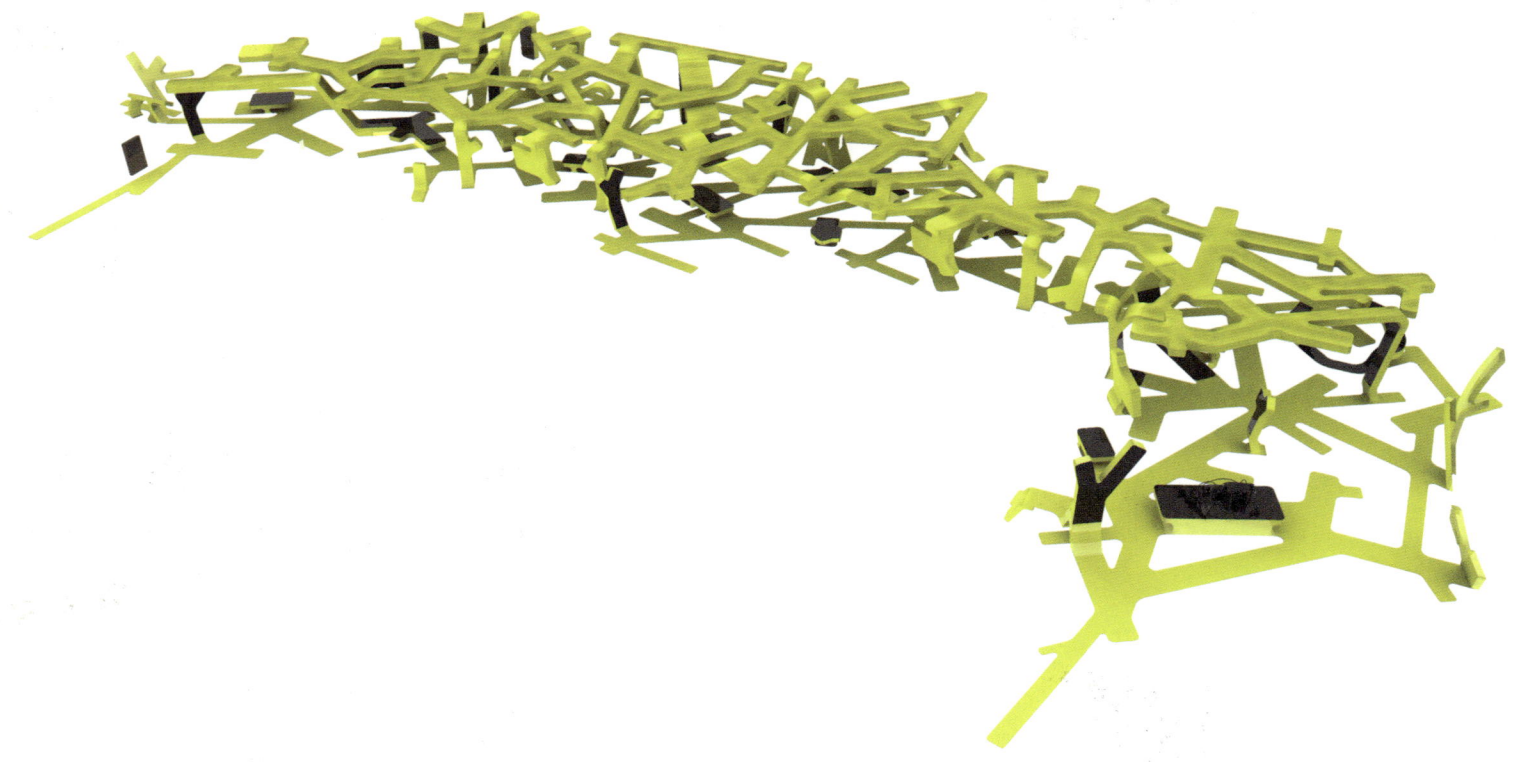

STRUCTURE

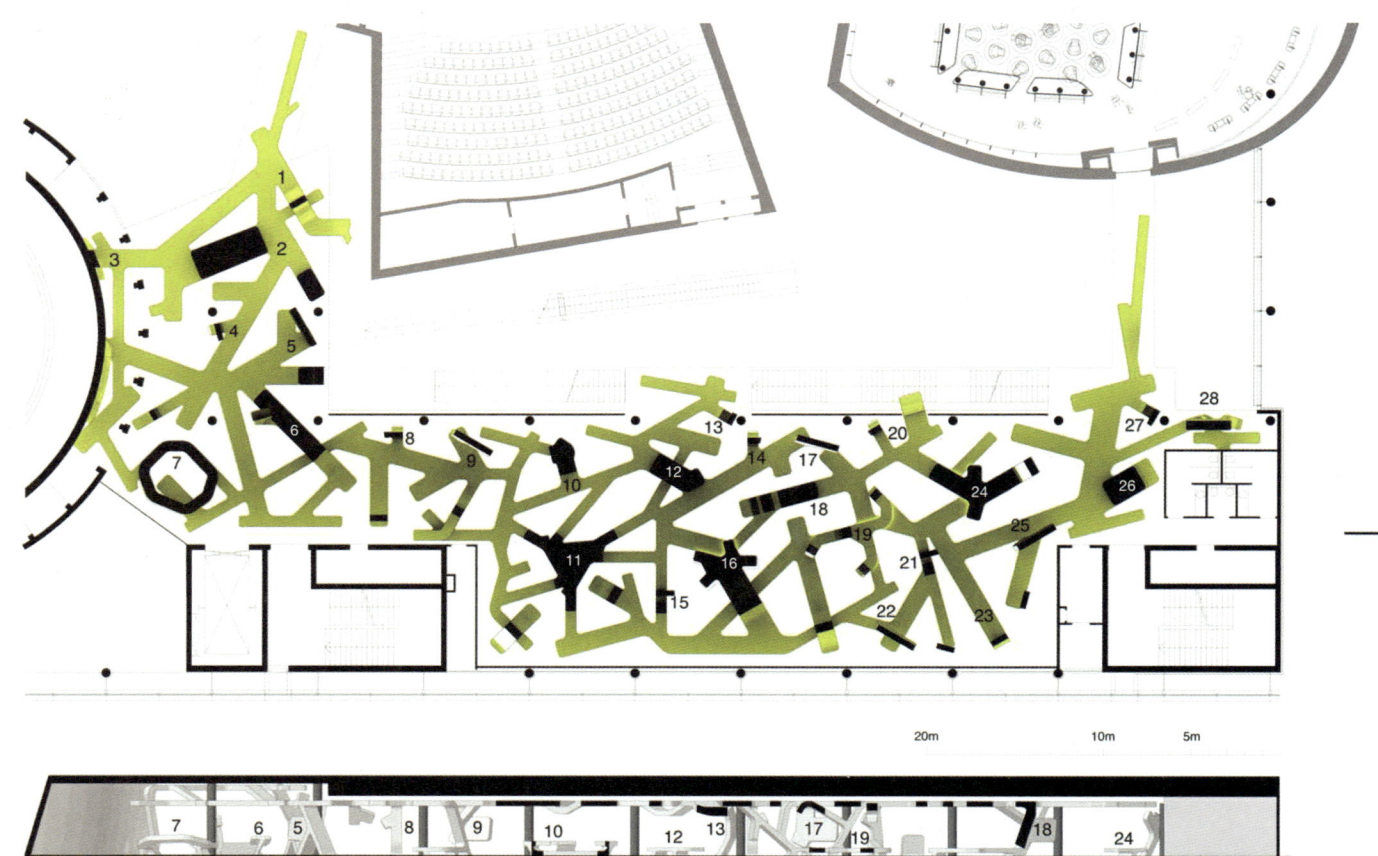

1 Haupteingang **2** Virtuelles Wasser **3** Wandvorlage 360° Kino **4** Verbrauch im Vergleich **5** Ökologischer Rucksack **6** Ökologischer Fußabdruck **7** Automotive Lifecycle
8 Nachhaltigkeit bei Volkswagen **9** Antriebs- und Kraftstoffstrategien **10** interaktive Stausimulation **11** Idee der Nachhaltigkeit **12** Mobilitätsprofile **13** Eingang Mitte
14 Mobilität der Zukunft **15** Gesellschaftliche Verantwortung **16** Brunnen der Erkenntnis **17** Utopientester **18** Exponat Klimawechselwirkungen **19** Klimazeiten
20 Bereichkennzeichnung Klimawechselwirkungen **21** Nachhaltigkeit und Ökonomie **22** nachhaltig Wirtschaften **23** Talking Head **24** Klimatexturen **25** Medienwand
Klimawechselwirkungen   **26** CO^2-Einsparpotentiale   **27**   Eingang Mobiglobe   **28**   Medienecke

FLOOR PLAN

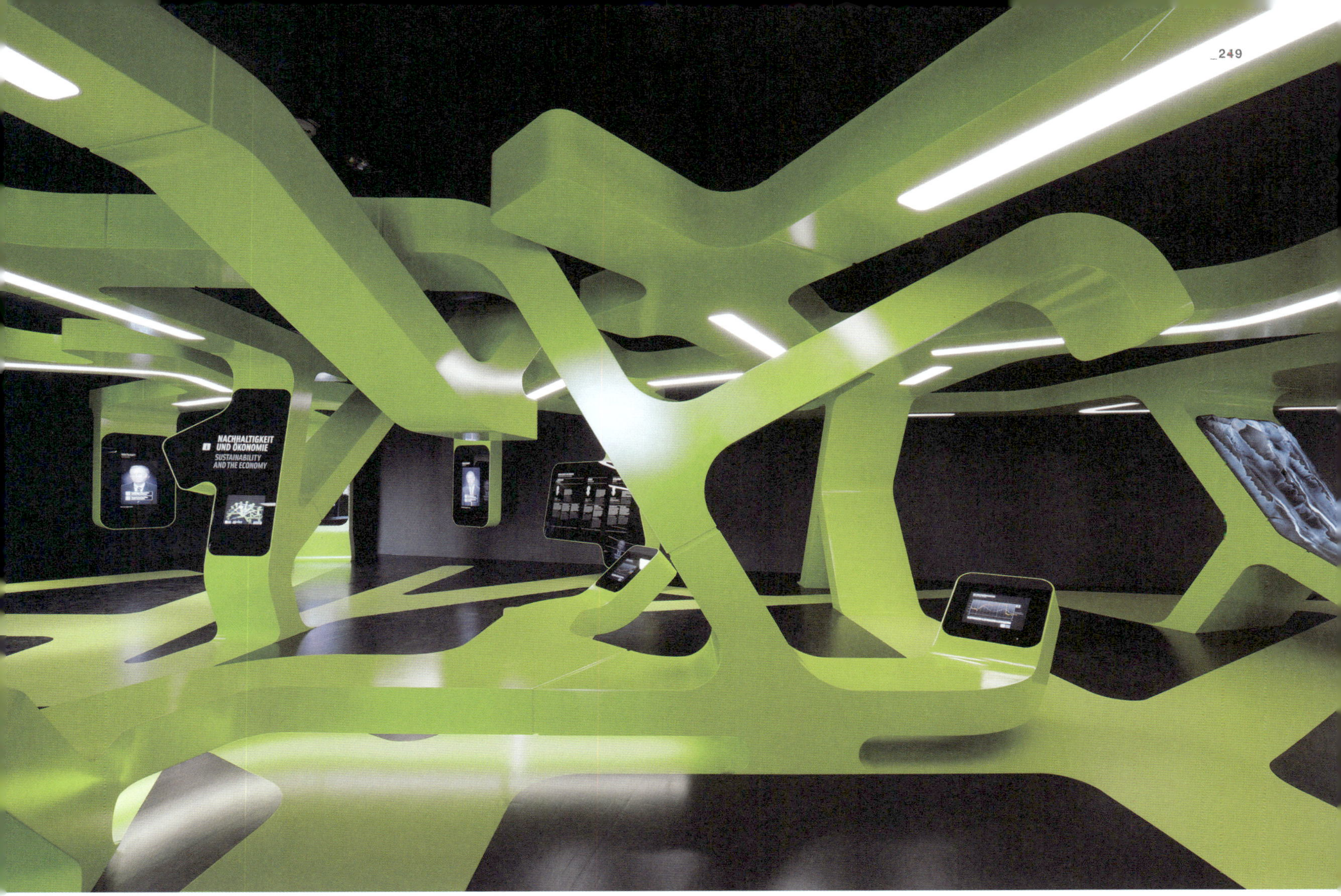
NACHHALTIGKEIT
UND ÖKONOMIE
SUSTAINABILITY
AND THE ECONOMY

LEVEL GREEN
Die Idee der Nachhaltigkeit
The concept of sustainability

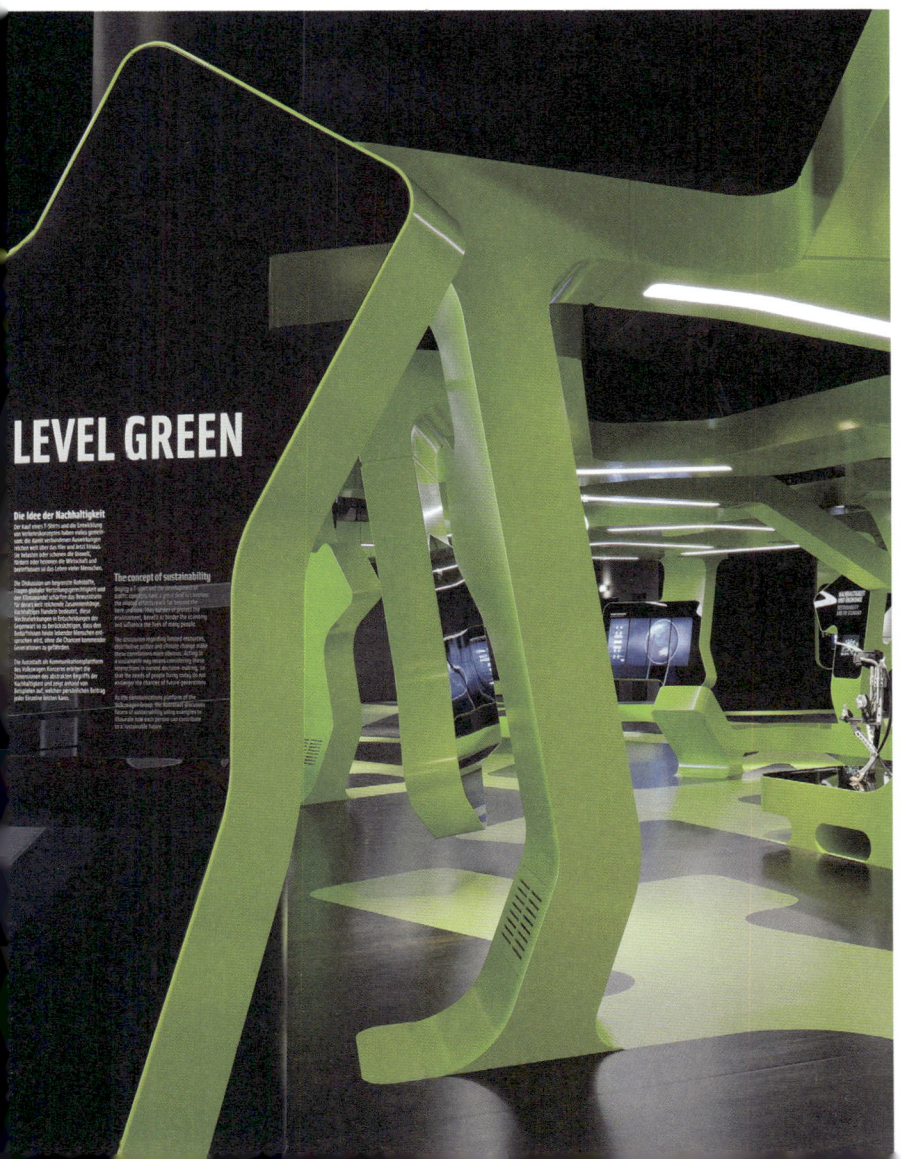

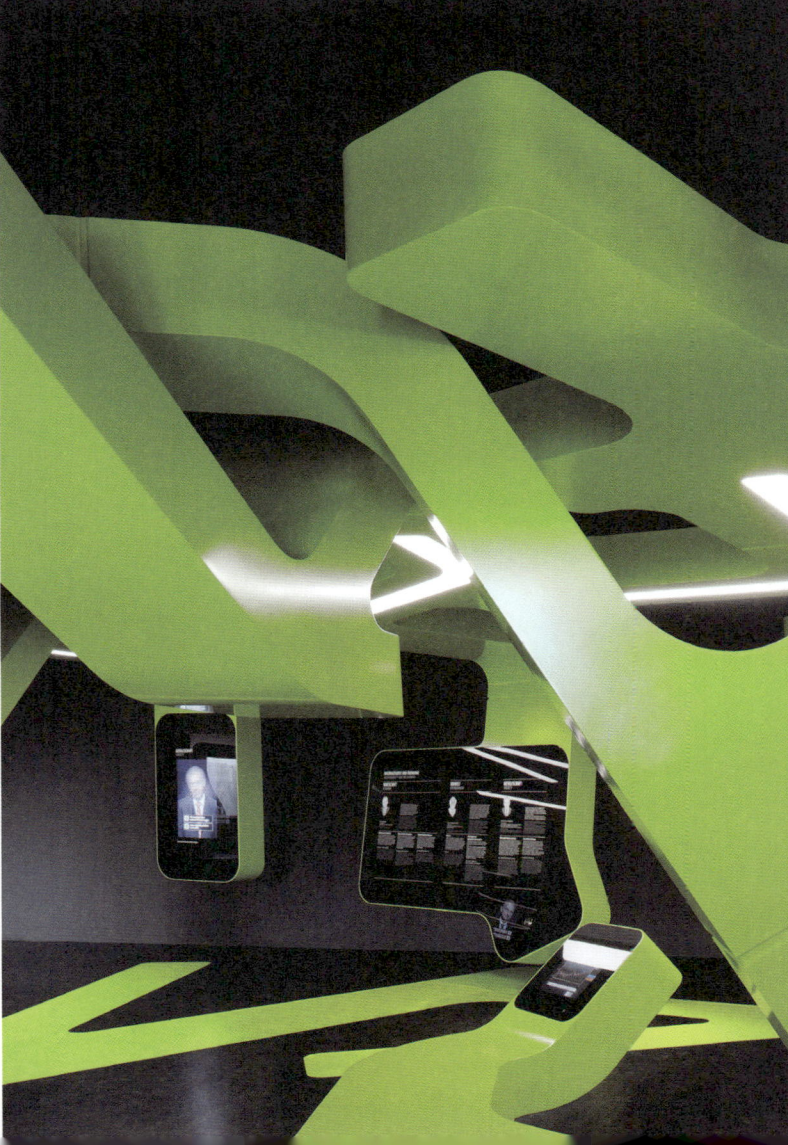

M.
OVER.
WALL

DESIGN_	J. MAYER H. ARCHITECTS
DESIGN TEAM_	JÜRGEN MAYER H., SEBASTIAN FINCKH, STEPHANIE KALLAENE
LOCATION_	BERLIN-MITTE, GERMANY
PHOTOGRAPHY_	J. MAYER H. ARCHITECTS
TOTAL FLOOR AREA_	APPROX. 1,200M²
PHASE_	PROJECT 2008-2009	COMPLETION 2010

M.OVER.WALL
INTERIER DESIGN FOR A PENTHOUSE

A residence of a young art-loving family in the middle of Berlin, independent from any conventional floor plan, with separate office and guest area on a continuous level and two additional penthouses. The premises articulate in a stretched and angled main space without functional determination and several adjacent subsidiary rooms for retreat with individually assigned programming.

The central space as a 'space of potentiality' can be modulated by mobile, rollable furniture elements with the ability to interlock in order to form flexible zones of use. It accommodates (customizes itself) like a universal cloth to the multiple, unforeseeable and continuously changing needs of the inhabitants. The palette reaches from a continuous exhibition space to a couple of private cabinets of graduated intimacy. The functional areas transform to a set of zones with more or less blurred transitions.

The traversing supergraphics reflect the image of floor and ceiling mirror-inverted (horizontal symmetry) and builds up a further abstraction of spatial qualities. Crucial for this is the colour concept with modulated yellowish greens and a complementary aubergine hue. All built-in wall closets are understood as integral part of the coloured walls, which become legible as operable fronts only by their premium quality surface.

With the constantly changing light of the wandering sun, which enters through the afloat glass façade, the colour and spatial impression of the graphic strings transmute continuously. The central living area in this subtle way becomes a visual laboratory and an imaginary space of itself.

PLAN

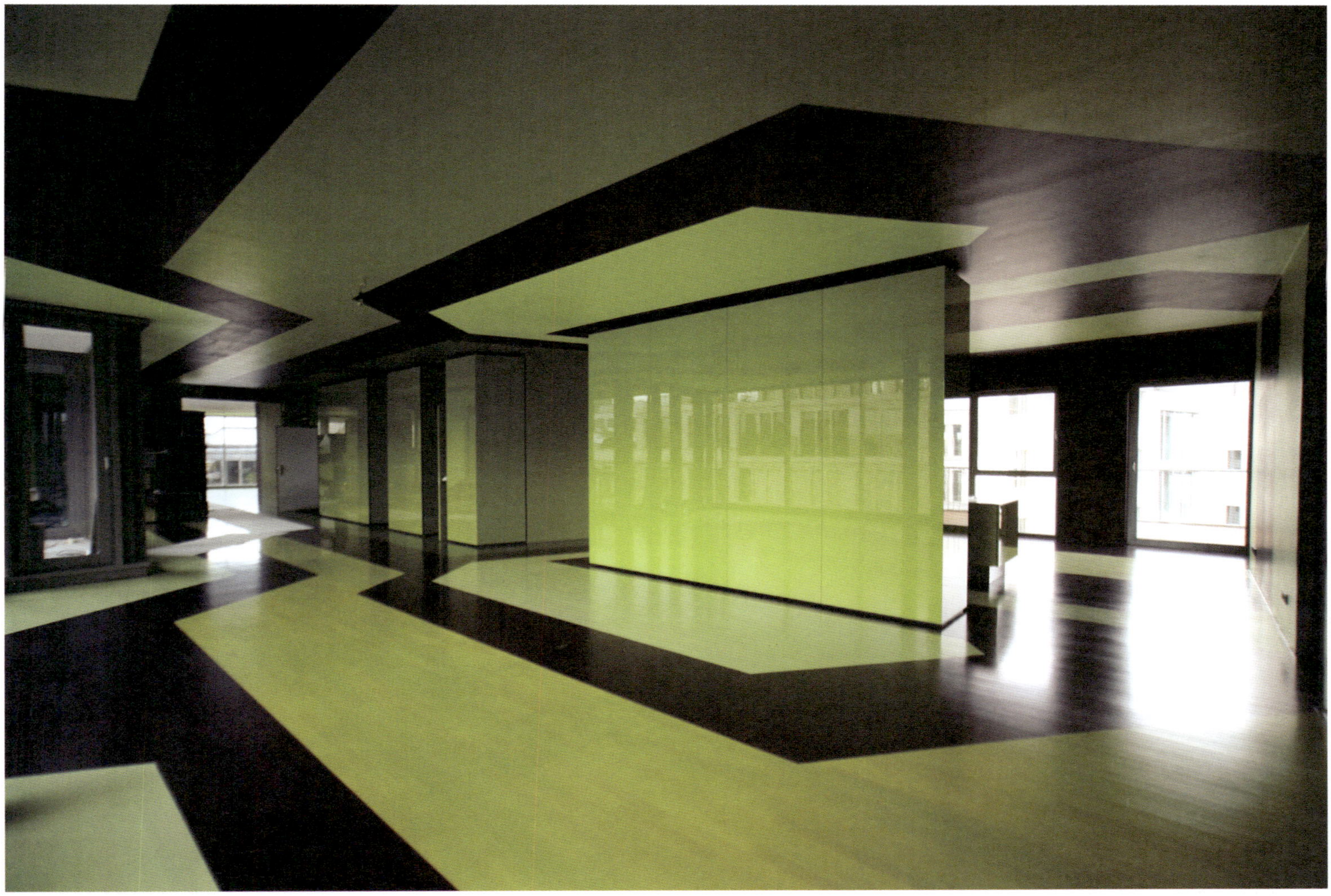

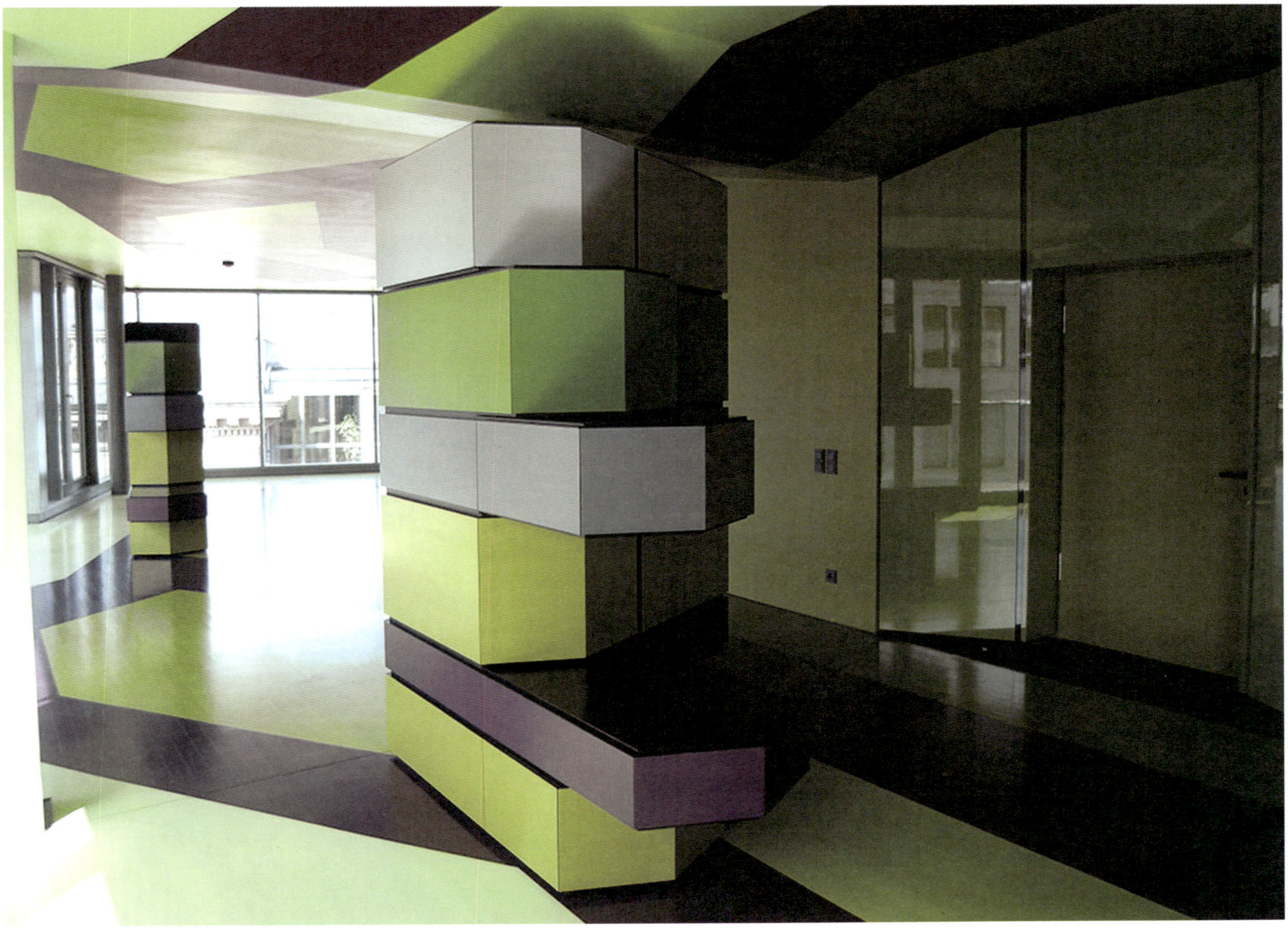

PRE.TEXT / VOR.WAND

DESIGN_ J. MAYER H. ARCHITECTS
DESIGN TEAM_ JÜRGEN MAYER H., WILKO HOFFMANN
LOCATION_ ZÜRICH & PRAGUE
PHOTOGRAPHY_ MAGNUS MÜLLER GALLERY
PROJECT_ 2010

PRE.TEXT / VOR.WAND
DATAPROTECTION PATTERN IN SHINGLE STYLE

Various lines of demarcation, or rather 'façades of countenance', have always separated the personal and the public. And in the case of information, the relationship between public and private becomes a complicated set of liabilities. It's a contract of confidentiality. By the beginning of the 20th century, information control generated a visual pattern called Data Protection Pattern or DPP that helps to veil personal information in print media. Letter and numbers, ingredients of information construction, are used in excess to create a speechless and slurry form of covering text.
Today, a new global network of unsecured data transfer remains to be resolved.
While DPP continues to proliferate in print media, it provides the model for carriers of electronic information, which are physically erased by overwriting the entire data carrier, or at least the used sectors, with a confusion of pattern.
An excess of information transforms the 'private' into apparent nonexistence.

GALERIE JAROSLAVA FRAGNERA, PRAGUE

MUSEUM FÜR GESTALTURG, ZÜRICH

GALERIE JAROSLAVA FRAGNERA, PRAGUE

DUPLI. CASA

DESIGN_	J. MAYER H. ARCHITECTS
DESIGN TEAM_	JÜRGEN MAYER H., GEORG SCHMIDTHALS, THORSTEN BLATTER, SIMON TAKASAKI, ANDRE SANTER, SEBASTIAN FINCKH
LOCATION_	MARBACH, GERMANY
CLIENT_	PRIVATE
PHOTOGRAPHY_	DAVID FRANCK
CONSTRUCTION AREA_	SITE 6,900M², BUILDING 569M², TOTAL FLOOR 1,190M²
HEIGHT_	12.20M
STRUCTURE_	FERROCONCRETE, BRICK, ROOF: TIMBER
PRINCIPAL EXTERIOR MATERIALS_	AERATED CONCRETE WITH PLASTER, GLASS
PRINCIPAL INTERIOR MATERIALS_	WOOD PANELLING, PLASTER, PARQUET FLOORING
PHASE_	PROJECT 2005-2007	COMPLETION 2008

DUPLI.CASA

The geometry of the building is based on the footprint of the house that was previously located on the site. Originally built in 1984 and with many extensions and modifications since then, the new building echoes the 'family archaeology' by duplication and rotation. Lifted up, it creates a semi-public space on ground level between two layers of discretion. The skin of the villa performs a sophisticated connection between inside and outside and offers spectacular views onto the old town of Marbach and the German national literature archive on the other side of the Neckar valley.

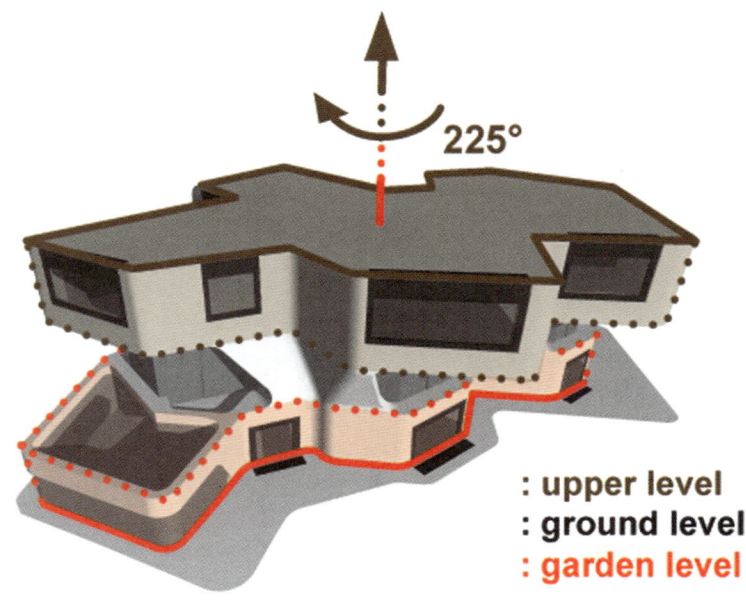

CONCEPT

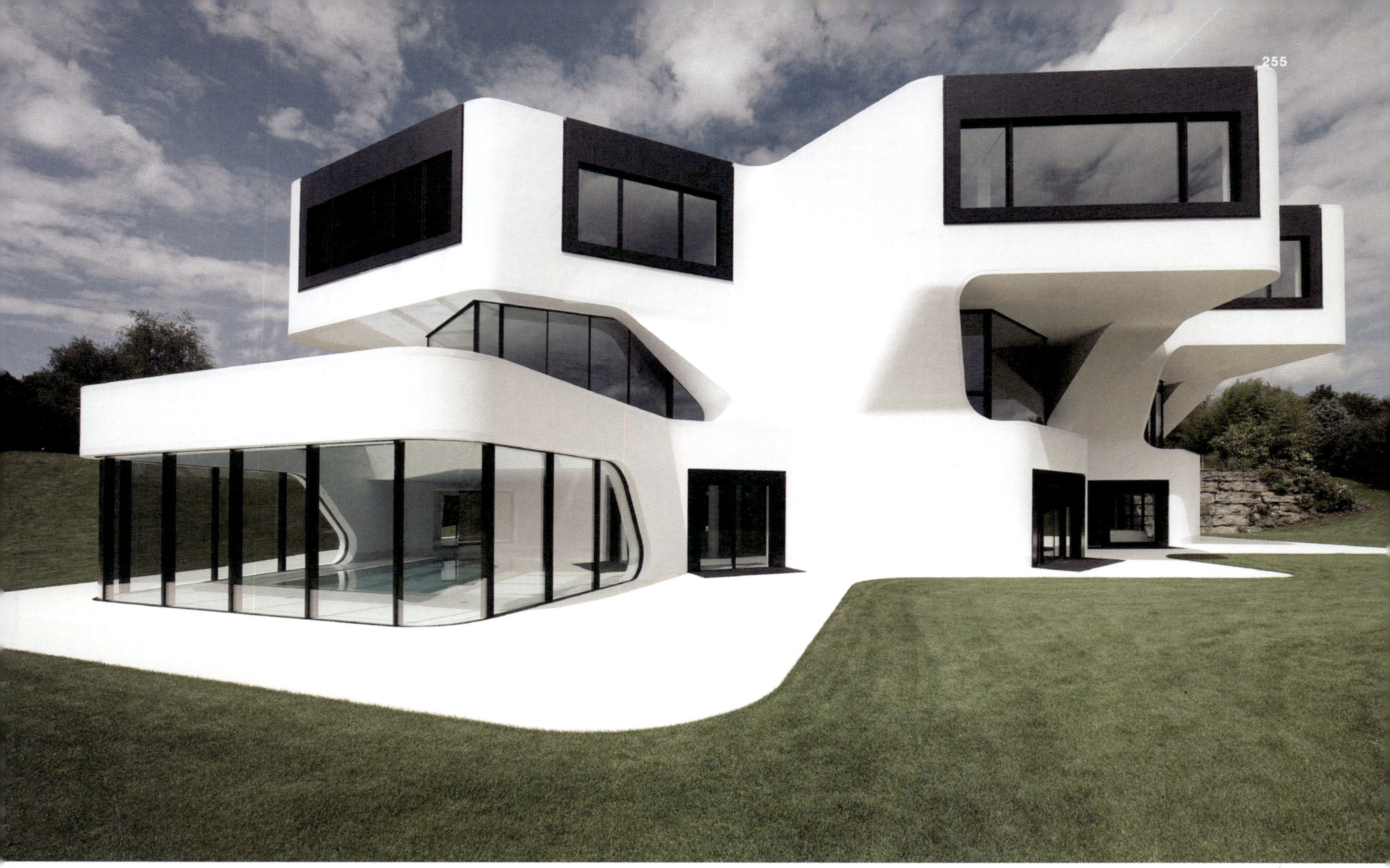

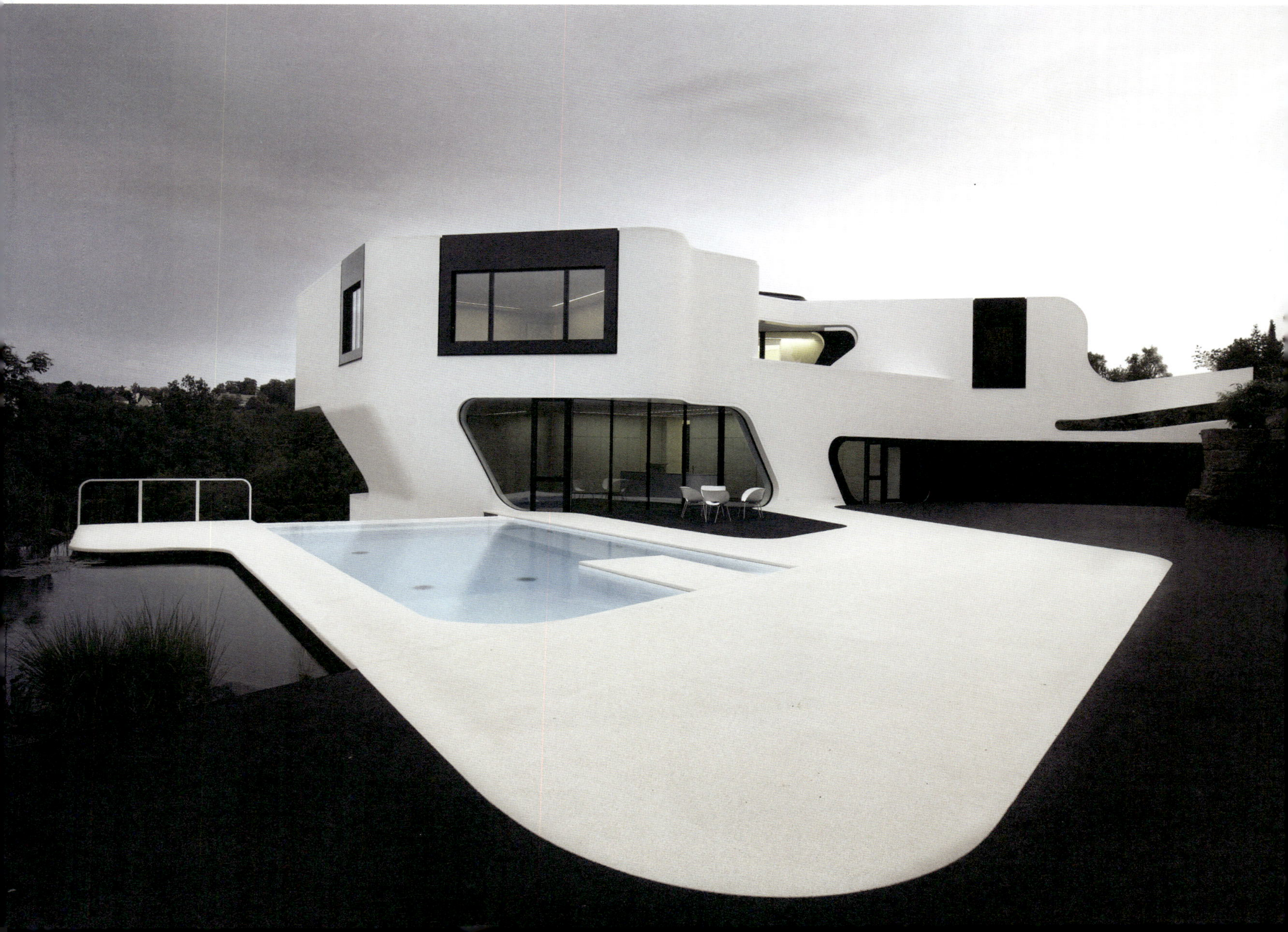

Contributors Index

3GATTI
Aterlier FCJZ
BBFL
bluarch architecture + interiors + lighting
Boris Kudlicka and WWAA
Campaign
Carsten Jörgensen
CCS Architecture
Chikara Ohno / sinato
Chrystalline Artchitect
Collaborative Architecture
design spirits co., ltd.
Dune
Elding Oscarson
emmanuelle moureaux architecture + design
Emmanuel Picault, Ludwig Godefroy
Facet Studio
FLATarchitects
hashimoto yukio design studio inc.
i29 l interior architects
Ippolito Fleitz Group
IwamotoScott Architecture
Jakob + MacFarlane Architects
J. Mayer H. Architects
Jonas Wagell Design & Architecture
Koan + / Koan Design
Koichi Takada Architects
Kotaro HORIUCHI
LABOR13
MAEZM + Sarah Kim
Ministry of Design
NAU, DGJ
Nuca Studio
Paradox Studio
Pedro Pacheco
RDAI
Sander Architecten bv
Straight Square Design
Studio MK27
Studio SKLIM
Suppose design office l Makoto Tanijiri
Teun Fleskens
The Metrics
Urban A&O, Thinc Design, Local Projects
Van Eijk & Van der Lubbe, Philips Design
WWAA + 137Kilo
Yoichi Yamamoto Architects
Zaha Hadid